Carolyn

Stephen

HERKIMER

AND THE

Darren

STAT PACK

Brenda

VENTURE INTO

Valarie

MONEY MATHEMATICS

Wayne

Glen

Janice

Roger

Frances

A series of vignettes involving ten students and their learning experiences relating to the mathematics of money management and financial literacy.

SANDERSON M. SMITH

AuthorHouse™
1663 Liberty Drive
Bloomington, IN 47403
www.authorhouse.com
Phone: 1-800-839-8640

First published by AuthorHouse 9/30/2009

ISBN: 978-1-4490-1905-1 (e)
ISBN: 978-1-4490-1906-8 (sc)

Library of Congress Control Number: 2009908495

Printed in the United States of America
Bloomington, Indiana

This book is printed on acid-free paper.

This book is dedicated to my brothers,

Ephraim K. Smith, Jr. (Koch)

and

Lawrence K. Smith (Larry)

and to the memory of our wonderful parents,

Katherine and Ephraim

TABLE OF CONTENTS

INTRODUCTION   8

Session #1: HERKIMER AND THE STAT PACK GET BACK TOGETHER   15

Session #2: INTRODUCTORY DISCUSSIONS ABOUT FINANCE   19

Session #3: SINGLE DEPOSIT ACCUMULATION   22

Session #4: SINGLE PAYMENT DISCOUNTING   28

Session #5: TRAPPING A FINANCIAL SOLUTION   34

HERKIMER'S QUICK QUIZ (Sessions 1 - 5)   39

Session #6: PACK MEMBERS CREATE USEFUL FINANCIAL TABLES   40

Session #7: THE CONCEPT OF FINANCIAL INFLATION   45

Session #8: MULTIPLE DEPOSIT ACCUMULATION   49

Session #9: FINANCIAL SITUATIONS INVOLVING PAYMENT ACCUMULATIONS   55

Session #10: OTHER MULTIPLE DEPOSIT ACCUMULATION MODELS   59

HERKIMER'S QUICK QUIZ (Sessions 6 - 10)   63

Session #11: SOME DEPOSITS DON'T FIT A NICE
ACCUMULATION MODEL   64

Session #12: PRESENT VALUE OF FUTURE
PAYMENTS   68

Session #13: MORE ON CALCULATING
INVESTMENT YIELDS   72

Session #14: THE PACK PRODUCES MORE USEFUL
FINANCIAL TABLES   76

Session #15: OH! THAT EVER-USEFUL ALGEBRA   81

HERKIMER'S QUICK QUIZ (Sessions 11 - 15)   85

Session #16: THE ANATOMY OF A LOAN   86

Session #17: INSTALLMENT PAYMENT
FORMULAS ARE SOMETIMES USEFUL   92

Session #18: DON'T CONFUSE NOMINAL RATES
AND TRUE RATES   97

Session #19: MORE ON NOMINAL RATES   102

Session #20: PAYMENTS AND INTEREST PERIODS
MUST JIVE   107

HERKIMER'S QUICK QUIZ (Sessions 16 - 20)   112

Session #21: LOANS WITH MONTHLY PAYMENTS   113

Session #22: MORTGAGES - THE BIG LOANS   121

Session #23: MORTGAGES - COMPLETE PAYMENT
              SCHEDULES   127

Session #24: THE CREDIT CARD TRAP   133

Session #25: CREDIT - A DANGEROUS WAY TO
              LIVE BEYOND YOUR MEANS   140

HERKIMER'S QUICK QUIZ (Sessions 21 - 25)   145

Session #26: BASIC CONCEPT BEHIND
              RETIREMENT PLANS   146

Session #27: A BIT MORE ON RETIREMENT PLANS   152

Session #28: SOME INSIGHT INTO BONDS   157

Session #29: YIELDS ON U.S. SAVINGS BONDS   161

Session #30: YIELDS ON COUPON BONDS   167

HERKIMER'S QUICK QUIZ (Sessions 26 - 30)   171

Session #31: AVOID PONZI SCHEMES   172

Session #32: PACK MEMBERS DO INDIVIDUAL
              PROJECTS   177

Session #33: BRENDA'S PROJECT: THE RULE OF 72   180

Session #34: VALARIE'S PROJECT: PAYDAY LOANS   185

Session #35: WAYNE'S PROJECT: START TO SAVE EARLY   189

Session #36: DARRIN'S PROJECT: CONTINUOUS INTEREST   194

Session #37: FRANCES' PROJECT: ADVANTAGE OF EXTRA MORTGAGE PAYMENTS   199

Session #38: CAROLYN'S PROJECT: REFINANCING   205

Session #39: ROGER'S PROJECT: BOND AMORTIZATION SCHEDULES   210

Session #40: STEPHEN'S PROJECT: INFLATION CONCERNS IN RETIREMENT   216

Session #41: GLEN'S PROJECT: TEASER RATES   222

Session #42: JANICE'S PROJECT; ADJUSTABLE RATE MORTGAGES (ARMs)   227

Session #43: THE PACK MAKES A LIST OF USEFUL FINANCIAL TABLES   233

Session #44: HERKIMER AND THE STAT PACK: THE FINAL SESSION   241

HERKIMER'S FINAL FINANCIAL TEST   246

APPENDIXES   256

APPENDIX A (Algebraic Derivation of Formulas)   257

APPENDIX B (A complete 30-year mortgage schedule
and a complete credit card payment schedule)   260

APPENDIX C (Solutions and comments for selected
activities)   271

INDEX OF TERMS (By session number)   294

INTRODUCTION
(Hey, why did I write this book?)

It was not the financial panic of 2008 that prompted me to begin this project. I started putting this book together prior to the start of the downhill ride into financial chaos for world economies. Yes, a few years ago I anticipated that it was only a matter of time before financial difficulties would become a reality for many families who were living well beyond their means due to the deadly combination of easily-available money and a lack of a basic understanding of how *money works*, but I did not anticipate the magnitude of the crisis that developed in the United States and worldwide.

Since my early years as a mathematics teacher I have been aware of the absence of basic financial mathematics education in our nation's educational system. As a mathematics major in college I took many advanced math courses, but none really related to using mathematics for practical money management in my adult after-college life. Practically all of my knowledge of financial mathematics came during a two-year period when I was an actuarial trainee with Travelers Insurance in Hartford, Connecticut. This represents a classic example of how real life experiences can be far more valuable than classroom learning. I was in my mid-twenties when I really learned how money works and gained an appreciation of the fact that I had control of my financial destiny. I did not have to rely on others to produce financial computations relating to my present and future money planning. Even in those pre-computer days I could use the primitive calculators that existed then and do financial computations involving personal finances. I had the **math power** to check on figures provided to me by so-called financial advisors. In a few instances I was able to establish that these "experts" were generating incorrect results with whatever methods (usually consulting numerical tables) they were using. Clearly, developing an understanding of money mathematics has been a major reason why I have never had financial problems. However, I fully realize that knowledge doesn't guarantee success. One is at least a partial product of childhood experiences and upbringing. Lessons learned in early life are important.

I often reflect on my childhood years growing up in a depressed coal-mining region in northeastern Pennsylvania. These were pre-credit card days. If one didn't have enough money to purchase an item, one simply had to "save up" until the needed amount was available. I am grateful that I learned the power of saving and the wisdom to totally avoid the dangers of the "take it now and pay for it later" mentality. I was fortunate that my wonderful and caring blue-collar parents had financial savvy in the sense that they never felt the necessity to have what everyone else had. They provided me and my two brothers with money wisdom that could never be found in formal education. Although our family had little money, saving "just a bit" every month was part of our home education. The thought of simply charging an item to a credit card and taking it home was beyond our imagination. If we wanted something our friends had we knew what we had to do to get it. I distinctly remember wanting a specific item some of my buddies had, but after saving enough money to get it I found that I really didn't want it anymore. I wonder what percent of impulsive credit card purchases represent buying items that one could easily do without. Many Americans were good savers until around 1980 when the credit card mentality and too-easy-to-get mortgage loans pretty much took over sound financial planning. The basic question seemed to shift from "do I have enough money to buy the item I want?" to "will I be able to make the minimum payment required on my credit card if I buy the item?" In my humble opinion, this simple question shift led many good people on the path to financial disaster. Saving for the future is difficult when one must use a good portion of each paycheck to pay for items purchased without cash sometime in the not-too-distant past.

Don't get me wrong. I'm not anti-credit card nor do I believe the credit card companies are the villains they are portrayed to be in much of the media. Credit card interest rates are high, but there is a very logical reason for this. We just need to know the consequences of making the minimal monthly payments we are allowed to make. If one can't figure this out, it is not the fault of the credit card company. And, there aren't too many lending organizations that provide an opportunity for an interest-free loan for a month, as credit card companies do. We don't need elaborate laws and regulations to protect us from financial demons who want to run us into perpetual debt. We simply need to understand how money works and how to manage personal finances to avoid future money problems. Given that we now have multitudes of adults who have demonstrated an inability to handle personal finances it is not likely that they will pass money management skills on to their children. Even with this unpleasant scenario, we should make efforts to improve financial literacy for all citizens, particularly young members of our capitalistic society.

During most of my 40 years of full time teaching at Cate School in Carpinteria, California I taught a **non-required** semester course titled *Mathematics of Finance*. The course was primarily designed for seniors who where not mathematically inclined and one of the major purposes was to show students how money works in a capitalistic society. It went far beyond simply teaching young people how to balance a checkbook. (Side note: Even today, I have a small number of intelligent adult friends who joke about their inability to balance their checkbooks. It's almost a sign of honor to make such an admission. I wonder how many people who can't read would take pride in acknowledging this shortcoming.)

During recent years advances in technology have made computational tasks considerably easier. If one understands basic algebraic principles, spreadsheets represent an incredible teaching tool. Using spreadsheets for computational work and graphic displays greatly enhanced the appeal of *Mathematics of Finance* at Cate School. Students learned the dangers of making minimal payments on credit card purchases. And, they learned how to produce a complete loan repayment schedule for a mortgage associated with a home loan. Eventually some seniors and juniors who were taking college-level math courses such as Advanced Placement Calculus and Advanced Placement Statistics enrolled in the course. In a *Cate School Bulletin* (Fall, 2007) featuring short writings from graduates relating to mathematics at Cate many referenced the non-required *Mathematics of Finance* course as an extremely valuable educational experience. Even today, five years into retirement from full-time mathematics teaching at Cate, I receive correspondence (mostly emails) from former students appreciative of the financial education provided in *Mathematics of Finance.*

During my final years at Cate I decided that in retirement I would write educational books on two very practical and useful mathematical subjects, *statistics* and *finance.* These would not be textbooks but rather a series of easy-to-read vignettes that would involve conversations and interactions between a fantasy cartoon character Herkimer and ten students who would call themselves the **Stat Pack.**

Now Herkimer has been my buddy for a long time and he is well known to about 30 years of Cate School graduates. He is definitely not a new face to many good people living all over the world. He's an interesting character whose eyes are never open and whose mouth is never closed. That's how he was "born" at Cate School. And that's how he will always remain.

The students, artistically created by Cate School graduate Verena Chu, would have names and artistic character recognition.

| Brenda | Carolyn | Darren | Frances | Glen |

| Janice | Roger | Stephen | Valarie | Wayne |

This group of ten students adopted the name **Stat Pack** when they were studying statistics and Herkimer was their "academic" advisor. Herkimer was not intended to be the teacher but rather would simply be present when Stat Pack members needed insight and guidance during their educational journey through *statistics* and *finance.* He would be visible only to Pack students who would engage in conversations with him and who would produce lots of chalkboard handwriting as they developed an understanding and appreciation of *statistical* and *financial* mathematics. Herkimer's role was NOT to be a pure lecture-type educator. He was an activity man who wanted students to do more than just listen. It was my intent to use this fantasy world atmosphere to make the reader's learning process a bit more entertaining then the reading one encounters in standard textbooks.

I decided to do the *statistics* book first. In that fantasy educational adventure the students involved adopted the name Stat Pack. The group made its way through this important and fascinating subject in my book THE STATISTICAL ODYSSEY OF HERKIMER AND THE STAT PACK. They then had a desire to move on to a study of *financial mathematics*.

It's important to understand that this, the *finance* book, is not meant to be an investment strategy guide. Herkimer and the Pack do not discuss stock market investment strategies, for instance. Rather, the book emphasizes use of the mathematics of finance. Among many other things, the reader will, like the Pack, learn how to use mathematics to accumulate and discount money payments, to illustrate the financial problems one can encounter by making minimum payments on credit cards, how to compute the interest portions of mortgage payments, why numerically-equal interest rates can differ considerably, and the inner workings of a Ponzi scheme.

As previously stated, my *finance* students at Cate School made extensive use of spreadsheets. In a similar manner, the Stat Pack will also do this. Spreadsheets and the related graphics are relatively easy to construct and it was neither difficult nor time consuming to get Cate students to use them effectively. However, it is NOT a purpose of this book to teach spreadsheet construction. In general, one would not have any reason to show the column labels (A, B, C, ...) and the row labels (1,2, 3, ...) and the "ideal" spreadsheet output might appear as in the following display:

Accumulation of daily earnings during a five-day week							
Day	Monday	Tuesday	Wednesday	Thursday	Friday		
Daily Earnings	$80	$120	$90	$60	$150		
TOTAL earnings	$80	$200	$290	$350	$500		

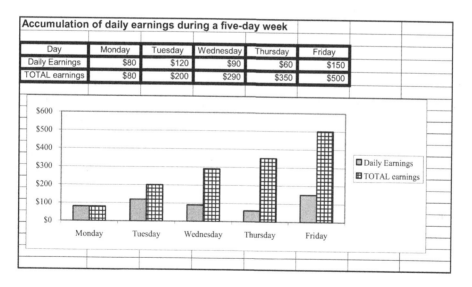

However, for most of the spreadsheet displays produced by Herkimer and the students the spreadsheet display will show row and column headings, as illustrated in the next illustration:

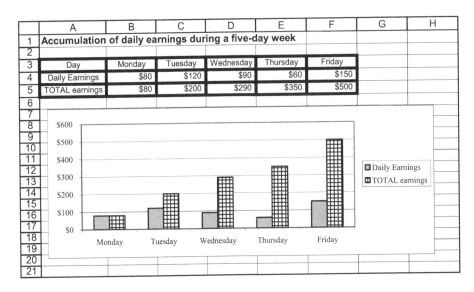

	A	B	C	D	E	F	G	H
1	Accumulation of daily earnings during a five-day week							
2								
3	Day	Monday	Tuesday	Wednesday	Thursday	Friday		
4	Daily Earnings	$80	$120	$90	$60	$150		
5	TOTAL earnings	$80	$200	$290	$350	$500		
6								
7								

This is done with the thought that one might possibly use this book to teach others the fundamental aspects of spreadsheets. For instance, if one is familiar with spreadsheet mathematics then the formula =**B5+C4** placed in cell C5 and **FILL**ed to the right does all the necessary computations.

One also needs to keep in mind that spreadsheet monetary figures are **formatted** to display two decimal places. After all, if a financial computation yields $54.6386512, who wants to look at all those digits after the decimal? Realistically this is just $54.64. Depending upon how calculations are done, this could result in minor differences in results obtained. For instance, the monthly payment on a 30-year $300,000 mortgage at 5.75% is $1,750.718569. (Don't fret! You will learn how this is calculated.) The total of the 30(12) = 360 payments is 360($1,750.718569) = $630,258.69 (with the total rounded to two decimal places). In spreadsheet arithmetic all decimals are used in computations although the results may, at the user's discretion, be rounded to two decimal places. Now the actual monthly payment rounded to two decimal places is $1,750.72. Using this rounded payment version the total of the payments, done on a calculator, would be 360($1,750.72) = $630,259.20. The difference of $0.51 is due to rounding. This is mentioned only to alert those who work on the activities that an answer obtained for a problem involving multiple payments might differ slightly from what is presented in the answer section.

While spreadsheet proficiency is not necessary to use this book, one who reads the book seriously and wishes to participate in suggested activities will need a calculator for requested computations. For instance, I doubt that most readers can do the computation

$$\$1000(1.065)^{20}$$

without calculator assistance. (Side note: With an indicated computation like this on the chalkboard I would tell my students that I could do it "in my head," but that I didn't want to "show off" and wanted them to do the calculator computation to make sure I had done it correctly. Once two or more students produced the same result I would write the result and say that the response "jived with mine.") A reader might wish to use developed formulas to check some monetary values appearing in displayed spreadsheets. Calculator efficiency is a valuable asset for this objective.

It is not a purpose here to teach the ***algebraic order of operations*** but knowledge of such is important if one (a) wants to make efficient use of a calculator for financial computations, and (b) desires to avoid serious miscalculations resulting from a lack of understanding of the hierarchy of basic number operations. With this in mind, the reader should know that the order of evaluation for indicated mathematical computations is:

> **First priority**: Parentheses and Brackets
> **Second priority**: Exponents
> **Third priority**: Multiplication and Division (from left to right)
> **Fourth priority**: Addition and Subtraction (from left to right)

Each session activity set will contain an ALCAL (ALgebra/CALculator) exercise providing the reader an opportunity to develop calculator efficiency relating to financial computations. Using * as the multiplications symbol and non-financial numerical expressions, here is a non-financial example:

	To be evaluated	Calculator entry	Number represented
(a)	$60-(5+8*2^2)/4*3$	$60-(5+8*2^2)/4*3$	32.25
(b)	$(60-5+8*2^2)/(4*3)$	$(60-5+8*2^2)/(4*3)$	7.25
(c)	$[60-5+(8*2)^2]/4*3$	$(60-5+(8*2)^2)/4*3$	233.25

Also, each session set will contain what is called a REFLECTION and COMMENT activity. In most instances these will be quotes that relate to (or can be related to) the world of finance and money. An individual reader might simply attempt to relate the quote to the session material. Groups might wish to discuss the quote in light of the session material. In any case, this activity exercise has no specific response or answer. When you encounter one of these activities I suggest that you simply "reflect" and "comment," especially in a situation when you have other interested friends in your vicinity.

I wish to express gratitude to my wife, Barbara, for being a partner in our journey to financial stability and for serving as a proof reader for this manuscript. I am also appreciative of the effort put forth by Steve Solano in assisting me with the placement of tables, graphs and images. Finally, I want to express gratitude to Cate School, my employer for 40 years, for offering me the opportunity to develop and teach a course related to financial literacy. As previously suggested there is evidence in the form of correspondence from former students that exposure to money mathematics was a valuable experience for students as they entered into the real world of business, finance and future planning.

Before moving into Herkimer's first session with the Stat Pack let me note that my parents were among many who lived through the Great Depression that started in the 1920's. Lessons were learned, the U.S. economy recovered, and opportunities and prosperity developed as citizens changed life styles and did those things necessary to get the country "started up" again. As I complete this book the financial crisis of the early 2000's is still present, but there are signs that the U.S. economy is starting on the long road to recovery. We can hope that the nation's youth are getting the wake up call and realize that they can't live in the fairy-tale financial world in which their parents have existed during the very early years of the 2000's. The real bottom line for youth is that you can't immediately have everything you want and you can't "keep up with the Joneses" by spending money that you don't have for items that you don't really need. And, perhaps the thought that everyone should note: **Every dollar you spend on the interest portion of a payment related to paying off financial debt is a dollar that can't be invested towards your future well-being.**

I attempted to write this book so that the casual reader can both enjoy it and benefit from it. For those who have a calculator, I do encourage use of it to work along with the Stat Pack as they do computations relating to their interactions with Herkimer. The mathematics of finance is not difficult to comprehend, but the required computations are not of the type that one can do mentally. Practice with efficient calculator use on complicated-looking numerical expressions can definitely pay future dividends and help one gain a solid appreciation for the power of basic algebra.

With these thoughts in mind let's start our fantasy adventure. I hope you enjoy Herkimer's journey on the path to financial literacy with the Stat Pack. While on this trail keep in mind that those who want to succeed will find a way to do so. Those who don't will find an excuse.

> *Persistent people begin their success at the point where others end in failure.*
>
> (Edward Eggleston)

Session #1
HERKIMER AND THE STAT PACK GET BACK TOGETHER

> *I must say I hate money, but it's the lack of it I hate most.*
> -Katherine Mansfield

It was a joyous reunion when Herkimer made his appearance before the Stat Pack. For those not familiar with the former association linking him to this group of young people, Herkimer had previous mathematical encounters with the ten students who were introduced in the book:

THE STATISTICAL ODYSSEY OF HERKIMER AND THE STAT PACK

Herkimer had provided guidance, advice, and educational activities as the students worked their way through an introductory statistics course. He loved the fact that the students had created a group name, the Stat Pack. When the Pack completed the statistics course they knew that Herkimer would disappear from their lives and that he would attempt to find and help another group of dedicated students who were about to undertake a study of basic statistics. Herkimer did not anticipate ever seeing the Pack again.

It was through association with Herkimer that the Pack came to realize that understanding the language of statistics was the real key to success in the subject. Herkimer often stressed the theme that *mathematics is a language.* He believed that far too many students fail to realize this simple fact and do nothing more than attempt to memorize procedures and formulas. Some Pack members reflected on previous math courses they had taken and realized that this is exactly what they had done. Herkimer had "opened their eyes" to the realization that learning mathematics involved much more than rote memorization. They came to appreciate that mathematics was indeed a language that had been developed over a period of 2000 years as humans struggled to understand and comprehend the universe in which they existed.

As they began thinking and talking about moving on to college and post-teen years, Pack discussions began to center around financial concerns. They realized that they would have to assume more responsibility for handling money and related items like credit cards. Always alert to current events, they were quite aware of the financial woes encountered by many adults. The time was early in the 21st century and Pack members, even at their young age, were aware that the United States was in the midst of a financial crisis. In fact, they knew the crisis was global. Terms such as *bankruptcy, meltdown, foreclosure, bailout,* and *bear market* dominated the news. And, they came to realize that they knew very little about how money works in a capitalistic society. While they all had studied statistics and some had taken a calculus course, they were provided no opportunities to take courses in financial mathematics and money management.

Will Rogers once said, "We are all ignorant, only in different subjects." As they were once ignorant about statistics, Pack members acknowledged that they were presently ignorant about finance. Some of them had listened to their parents moan over the fact that their mortgage payments had recently increased. Some knew people who were coming to grasp the reality of the fact that they were over their heads in credit card debt. The Pack knew that it was possible to buy items without paying cash and that it would be very easy to overestimate an ability to pay for these things when it came time to "pay up." But this group of youngsters realized that other than infrequent advice offered by parents they had no real experience with money management.

One of the group mentioned Herkimer and the statistics education they had received from him. Another wished aloud that Herkimer could somehow return to help them in developing an understanding of financial mathematics. This feeling was unanimous. The Pack had a genuine desire to gain knowledge and insight into how money works in a capitalistic society, but they were certain that they had seen the last of Herkimer.

They sat in silence for a brief period of time. Suddenly an eerie sense of restlessness gripped the group. For a brief moment they all knew something was about to happen.

At that moment, Herkimer reappeared.

When the student is ready, the teacher shall appear.
(Chinese proverb)

HERKIMER'S FASCINATING FINANCIAL FACTS:

Examine the serial numbers on the George Washington side of a one-dollar bill. Note that each starts with an alphabetical letter that is followed by eight number digits, and ends with a letter. Now take all the one-dollar bills you have in your possession and examine the eight digits in the serial numbers. It is highly unlikely that you will find an eight-digit sequence that does not contain a repeat digit.

Assuming the digits are selected by a random process, the probability of obtaining an eight-digit number with no repeat digit is $(10)(9)(8)(7)(6)(5)(4)(3)/10^8 = 0.018144$, or approximately 2%. Hence the probability that eight randomly-selected digits have at least one repeat digit is approximately 98%.

ACTIVITY SET FOR SESSION #1

1. **ALCAL** activity: The primary purpose in this activity is to emphasize an understanding of the *order of operations* as referenced in the INTRODUCTION. In each situation, evaluate the given numerical expression without any calculator use. Then use the calculator representation to check your responses:

	To be evaluated	Calculator entry	Number represented
(a)	5+10*2	5+10*2	
(b)	(5+10)*2	(5+10)*2	
(c)	$5+10^2$	5+10^2	
(d)	$(5+10)^2$	(5+10)^2	
(e)	$2*3^2$	2*3^2	
(f)	$(2*3)^2$	(2*3)^2	
(g)	100/5*2	100/5*2	
(h)	100/(5*2)	100/(5*2)	
(i)	100/2/2	100/2/2	
(j)	100/(2/2)	100/(2/2)	

2. REFLECTION & COMMENT activity:

   It is important for you to know where you are and where you want to go.

   (Common financial proverb)

3. Research the Internet for information relating to financial education. Using a search engine and typing in phrases such as *lack of financial education* will produce some interesting and eye-opening results.

4. There are numerous categories of financial education that are important to American citizens. Four of these are (a) basic savings; (b) credit management; (c) home

ownership; (d) retirement planning. Does your local school system provide education in these and other areas of finance?

5. Keeping in mind that some adults lack money management skills, what role should parents play in developing financial literacy for their children?

$$$
$$

Session #2
INTRODUCTORY DISCUSSION ABOUT FINANCE

> *When a fellow says it hain't the money but the principle o' the thing, it's th' money.*
> -Frank McKinney Hubbard

The Pack members were intelligent and curious students. They took time each day to keep up on current events. They were aware that a mortgage crisis was a major concern for the United States in the early years of the 21st Century. Always the optimist, Herkimer started out by saying that this crisis may well create momentum for improving financial literacy in the U.S. educational system. "Hey, it's about time," he said. "Too many people of all ages simply do not know how to manage money."

Herkimer went on to explain to the Pack that the need for financial knowledge has increased drastically in recent years. If one looks back 30 or 40 years, one finds people who shopped primarily with cash, had fixed-rate mortgages, and relied on company pensions for retirement. In modern times citizens of the United States have been introduced to a world of credit cards, adjustable rate mortgages, and a variety of pension problems along with uncertainty relating to the future of social security.

Herkimer also noted that some high schools offer courses that teach students how to balance a checkbook or follow the stock market, but only 18 states require some type of instruction in personal finance. Very few people know how an annuity works. And, it is also a fact that many highly educated people aren't financially literate and rely on so-called financial advisors for advice relating to investments and money matters.

"Financial literacy is a necessity in planning for retirement, for understanding how investment money grows, for managing debts, and for credit card management," said Herkimer as he began writing a question on the chalkboard. He mentioned a recent study indicating that only 18% of American adults questioned were able to produce a correct response to the question.

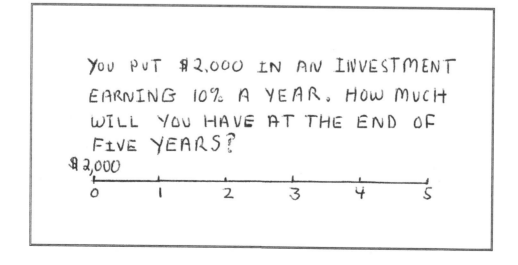

19

The challenge was before them. Pack members started punching numbers into their calculators. After about 30 seconds, Stephen asked, "I get $3,000, is that right?"

"No, responded a grinning Herkimer, "but that is the amount given by 34% of the adults who responded to the question. A sad fact is that 82% of the adults answering the question either got that answer, got another wrong answer, or simply didn't respond."

"I got $3,000 too," volunteered Glen. "Let's get started on gaining financial literacy," he continued. "It seems like we really need it."

HERKIMER'S FASCINATING FINANCIAL FACTS:

The Independence Hall clock is visible on the back of a $100 bill. While it is difficult to view without a magnifying class, the hands on the clock appear to be set at 4:10. It's not known if there is any particular significance to the displayed time.

Some have suggested that the time might actually be 2:21. What do you think?

ACTIVITY SET FOR SESSION #2

1. **ALCAL** activity: In each situation, evaluate the given numerical expression without any calculator use. Then use the calculator representation to check your responses:

	To be evaluated	Calculator entry	Number represented
(a)	$100-2*5^2+4/2^2$	100-2*5^2+4/2^2	
(b)	$88/4/2+3^3-(2*6/2)^2$	88/4/2+3^3-(2*6/2)^2	
(c)	$2^2*3^2*4^2$	2^2*3^2*4^2	
(d)	$(2*3*4)^2$	(2*3*4)^2	
(e)	3^2+4^2	3^2+4^2	
(g)	$(3+4)^2$	(3+4)^2	

2. REFLECTION & COMMENT activity:

It's not the government's role to save someone who is in debt because they went to the mall and charged thousands on their credit card. When many people budget, they start by thinking how to reduce their lattes or cable TV service. What's more important is to focus on the big expenses you need to pay each month, with about 50% of your income going toward that. Another 30 percent should could towards "flexible" costs including eating out, clothing, vacations and hobbies. Save the remaining 20 percent. That means consider housing needs before wants.

(Elizabeth Warren, Harvard Law Professor, "champion for consumers")

20

3. Can you provide the correct answer to the question posed by Herkimer in this vignette? (Answer will be provided in Session #3.)

4. Far too many Americans don't seem to understand the simple concept of not spending more than they earn. One major reason for this is that people think of credit cards as assets. Research the recent explosion of credit card use in the United States. Are credit card companies to blame for the considerable credit card debt in this country? Or, should the blame for this problem be attributed to a lack of basic financial education? (Sessions in this book will provide details relating to credit card use.)

5. What does it mean to say that a major cause of the financial turmoil in the early 2000's resulted from mortgage loans that were made to shaky borrowers? (Sessions in this book will discuss mortgage loans.)

$$$
$$$

Session #3
SINGLE DEPOSIT ACCUMULATION

> *The safest way to double your money is to fold it over once and put it in your pocket.*
> -Frank McKinney Hubbard

It was a new day but Herkimer quickly returned to the perplexing problem of the previous session. He placed this transparency on the overhead.

> **If you have $2,000 in an investment earning 10% a year and simply let the money grow, how much would you have at the end of 5 years?**

"OK," began Herkimer, "I bet those of you who came up with an answer of $3,000 probably calculated ($2,000)(0.10) = $200 interest per year and concluded that in 5 years the interest earned would be ($200)(5) = $1,000. Hence the value of their investment would be $2,000 + $1,000 = $3,000."

"Yup, that's what I did," responded Stephen.

"Me, too," chimed in Janice.

"Your responses are not without merit," said Herkimer. He explained that $3,000 would be a correct response if one is working with *simple interest*, where annual interest is calculated only on the initial investment. But Herkimer stressed that this is not the way money usually works in our capitalistic society. Simple interest does not take into account the power of *compounding*. Using compound interest, the $2,000 investment would accumulate to $2,000 + (0.10)($2,000) = $2,000(1 + 0.10) = $2,000(1.10) = $2,200 at the end of 1 year.

Always ready to emphasize the importance of basic algebra concepts, Herkimer placed this transparency on the overhead.

> **BASIC ALGEBRA**:
> a + ab = a(1+b).
> In general, ca + cb = c(a+b).

He went on to explain that one now starts the second year with $2,200. This new amount earns 10% annual interest. So, your investment value at the end of the second year is valued at $2,200 + (0.10)($2,200) = $2,200(1.10) = $2,240.

22

Herkimer wanted to emphasize the algebra involved and leading up to an important general formula so he slipped this transparency on the overhead below the one already there.

$$\$2{,}000(1.10) + (0.10)[\$2{,}000(1.10)]$$
$$= \$2{,}000(1.10)(1 + 0.10)$$
$$= \$2{,}000(1.10)(1.10)$$
$$= \$2{,}000(1.10)^2 = \$2{,}420$$

The total interest earned at the end of the second year is $2,240 - $2,000 = $240.

"Oh, I get it," bellowed Glen. The Pack watched as Glen took his calculator to the chalkboard and wrote the following:

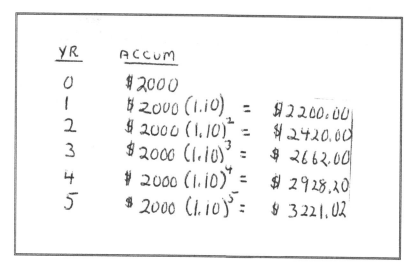

YR	ACCUM	
0	$2000	
1	$2000 (1.10)	= $2200.00
2	$2000 (1.10)²	= $2420.00
3	$2000 (1.10)³	= $2662.00
4	$2000 (1.10)⁴	= $2928.20
5	$2000 (1.10)⁵	= $3221.02

"You got it," replied Herkimer. "OK, now you guys learned how to use spreadsheets when we were studying statistics. Split up in pairs and see which team will be the first duo to produce a spreadsheet that will display the growth of single deposit of $2,000 for 10 years at both simple and compound annual rates of 10%." While column headings (A, B, C, ...) and row headings (1, 2, 3, ...) are often omitted when spreadsheets are printed out, Herkimer asked the Pack to include these on their displays. "After a few sessions we will not print out these headings," he continued, "but we will do so initially since they might be useful to a teacher who might want to use them to illustrate spreadsheet commands to students."

From their statistics experience the Pack knew how to determine five teams of two students through a random process. Once this was done, the teams moved to the available computers and started working on the project. After about 10 minutes the team of Carolyn and Wayne produced this display.

	A	B	C	D	E	F	C	H
1		**SIMPLE INTEREST**			**COMPOUND INTEREST**			
2	End Year	Interest Earned	Accum (10% simple interest)		Interest Earned	Accum (10% compound interest)		
3	**0**	$0.00	**$2,000.00**		$0.00	**$2,000.00**		
4	**1**	$200.00	**$2,200.00**		$200.00	**$2,200.00**		
5	**2**	$200.00	**$2,400.00**		$220.00	**$2,420.00**		
6	**3**	$200.00	**$2,600.00**		$242.00	**$2,662.00**		
7	**4**	$200.00	**$2,800.00**		$266.20	**$2,928.20**		
8	**5**	$200.00	**$3,000.00**		$292.82	**$3,221.02**		
9	**6**	$200.00	**$3,200.00**		$322.10	**$3,543.12**		
10	**7**	$200.00	**$3,400.00**		$354.31	**$3,897.43**		
11	**8**	$200.00	**$3,600.00**		$389.74	**$4,287.18**		
12	**9**	$200.00	**$3,800.00**		$428.72	**$4,715.90**		
13	**10**	$200.00	**$4,000.00**		$471.59	**$5,187.48**		
14								
15	TOTALS	**$2,000.00**			**$3,187.48**			

Chart (rows 16–33): Bar chart with x-axis years 0 through 10, y-axis $0 to $6,000. Legend: Accum (10% simple interest); Accum (10% compound interest).

Herkimer asked the Pack to use their calculators to check the computation displayed on the spreadsheet. It was agreed that at the end of the 10th year with compound interest, the accumulation would be $2000(1.10)^{10} = \$5187.48$. The interest earned would be $5187.48 - \$2000 = \$3187.48$.

GENERAL FORMULA FOR SINGLE DEPOSIT ACCUMULATION
(See Appendix A for algebraic derivation)

The accumulation of a single deposit of P for n years at annual interest rate i is
$$P(1+i)^n.$$

The total interest earned is
$$P(1+i)^n - P = P[(1+i)^n - 1].$$

"OK," said Herkimer. "As we move on, we are going to assume that we are working only with compound interest rate. This will be the first and last activity set to address the concept of simple interest. For now let's have all of you work up a spreadsheet relating to the accumulation of a single deposit.

Brenda came up with this spreadsheet display.

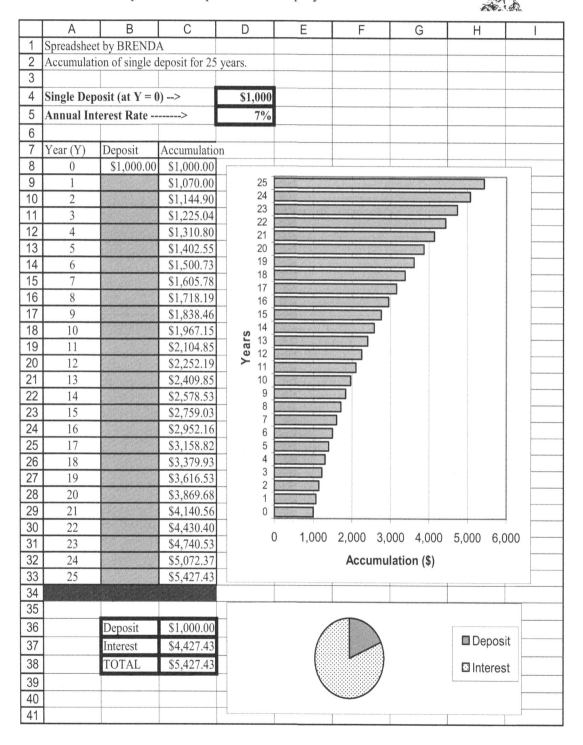

	A	B	C	D	E	F	G	H	I
1	Spreadsheet by BRENDA								
2	Accumulation of single deposit for 25 years.								
3									
4	Single Deposit (at Y = 0) -->			$1,000					
5	Annual Interest Rate -------->			7%					
6									
7	Year (Y)	Deposit	Accumulation						
8	0	$1,000.00	$1,000.00						
9	1		$1,070.00						
10	2		$1,144.90						
11	3		$1,225.04						
12	4		$1,310.80						
13	5		$1,402.55						
14	6		$1,500.73						
15	7		$1,605.78						
16	8		$1,718.19						
17	9		$1,838.46						
18	10		$1,967.15						
19	11		$2,104.85						
20	12		$2,252.19						
21	13		$2,409.85						
22	14		$2,578.53						
23	15		$2,759.03						
24	16		$2,952.16						
25	17		$3,158.82						
26	18		$3,379.93						
27	19		$3,616.53						
28	20		$3,869.68						
29	21		$4,140.56						
30	22		$4,430.40						
31	23		$4,740.53						
32	24		$5,072.37						
33	25		$5,427.43						
34									
35									
36		Deposit	$1,000.00						
37		Interest	$4,427.43						
38		TOTAL	$5,427.43						
39									
40									
41									

"That is a fabulous display," said Herkimer. "It clearly illustrates the power of compounding. Brenda's spreadsheet shows that at 7% money will double in 11 years, triple in 17 years, quadruple in 21 years, and increase 5-fold in 24 years. This should serve to illustrate that if you are young and are thinking about saving for retirement, the sooner you get started, the better."

HERKIMER'S FASCINATING FINANCIAL FACTS:

The United States Department of the Treasury first issued paper money in 1862 to make up for the shortage of coins and to finance the Civil War. During that period coins were made of gold and silver. Uncertainty due to the War caused people to hoard coins rather then spend them on items that fluctuated considerably in value over short spans of time.

ACTIVITY SET FOR SESSION #3

1. **ALCAL** activity: Use your calculator to check the interest earned total on Brenda's spreadsheet:

To be evaluated	Calculator entry	Number represented
$1000(1.07)^{25}$ - $1000	1000*1.07^25-1000	

2. REFLECTION & COMMENT activity:

   Unfortunately, we have found that the vast majority of people we deal with are more concerned with making money than with understanding how to make more efficient use of the money they already have.

   (*Money Mastery*, by Williams, Jeppson, and Botkin)

3. Assume someone unfamiliar with financial matters asks you to distinguish between simple interest and compound interest. How would you respond?

4. Consider a single deposit of $1,000. What would this investment be worth at the end of 20 years (a) at a simple interest rate of 5% a year? (b) at a simple interest rate of 10 % a year?

5. Consider a single deposit of $1,000. What would this investment be worth at the end of 20 years (a) at a compound interest rate of 5% a year? (b) at a compound interest rate of 10% a year?

6. Respond TRUE or FALSE.

(a) At a simple interest rate, a single deposit earning an annual rate of 10% will earn exactly twice as much interest as the same deposit earning an annual rate of 5% over a period of two or more years.

(b) At a compound interest rate, a single deposit earning an annual rate of 10% will earn exactly twice the interest as the same deposit earning an annual rate of 5% over a period of two or more years.

7. Since $\$1(1.05)^{14} = \$1.98$ and $\$1(1.05)^{15} = \$2.07$, money will double ($\$1$ will grow to $\$2$) in 15 years at annual compound interest rate of 5%. Complete the doubling-time table for the indicated annual compounded interest rates.

Rate	1%	2%	3%	4%	5%	6%	7%	8%	9%	10%
Years to double					**15**					

8. A single deposit of $\$5,000$ earns an annual rate of 6% for four years and an annual rate of 8% for the next six years. Assuming compound interest rates, what will be the value of this investment at the end of ten years?

$$
$$$

27

Session #4
SINGLE PAYMENT DISCOUNTING

> *Goodness is the only investment that never fails.*
> -Henry David Thoreau

The Pack gathered for another financial session. Never one to waste time, Herkimer put this transparency on the overhead while stating that from this point on all interest rates should be considered as compound rates.

> *Suppose you wanted to make a single deposit now and have it accumulate to $25,000 at the end of ten years. If the deposit can earn 10% per year, what amount x would be required to accomplish that goal?*
>
> | Deposit | $x | | | | | | | | | | $25,000. |
> |---|---|---|---|---|---|---|---|---|---|---|---|
> | End of Year | 0 | 1 | 2 | 3 | 4 | 5 | 6 | 7 | 8 | 9 | 10 |

Pack members definitely saw algebra here. Using the formula from the previous session, they realized the value of x would be obtained by solving the *linear equation* $x(1.10)^{10} = \$25,000 \implies x = \$25,000/(1.10)^{10} = \$25,000(1.10)^{-10} = \$9,638.58$.

BASIC ALGEBRA:
If a is not zero, then $ax = b \implies x = b/a$.

"Hey," said Glen. "How about asking how long it would take $5,000 to grow to $25,000 at a rate of 7%. We've got another algebraic situation that doesn't involve a linear equation."

The Pack readily agreed that the equation to be solved is $5000(1.07)^x = 25,000$. Wayne indicated that they were clearly in second-year algebra territory. He produced the following analysis on the chalkboard:

$$5000(1.07)^x = 25,000$$
$$\Rightarrow (1.07)^x = 5$$
$$\Rightarrow x \log(1.07) = \log(5)$$
$$\Rightarrow x = \log(5)/\log(1.07)$$
$$= 23.7876 \approx 24 \text{ (years)}$$

BASIC ALGEBRA:
If a and b are positive numbers, then $a^x = b \implies x = \log(b)/\log(a)$.

28

"OK," said Herkimer. "Let's now get into randomly selected teams of two and see who can produce a spreadsheet display involving both types of problems we just discussed."

The Pack loved spreadsheet challenges. It wasn't long before the team of Valarie and Stephen came up with this display:

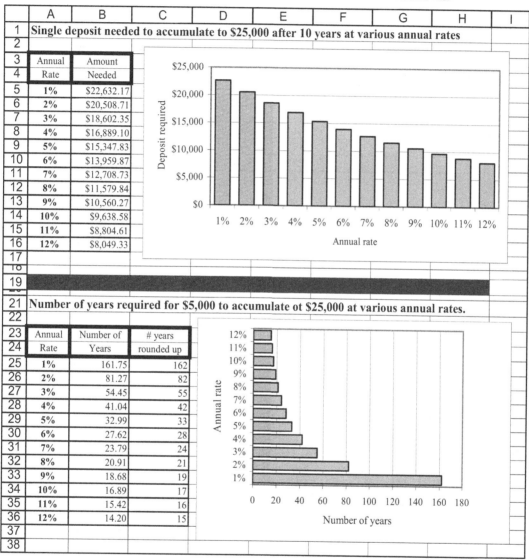

	A	B	C
1	Single deposit needed to accumulate to $25,000 after 10 years at various annual rates		
2			
3	Annual	Amount	
4	Rate	Needed	
5	1%	$22,632.17	
6	2%	$20,508.71	
7	3%	$18,602.35	
8	4%	$16,889.10	
9	5%	$15,347.83	
10	6%	$13,959.87	
11	7%	$12,708.73	
12	8%	$11,579.84	
13	9%	$10,560.27	
14	10%	$9,638.58	
15	11%	$8,804.61	
16	12%	$8,049.33	
17			
18			
19			
20			
21	Number of years required for $5,000 to accumulate ot $25,000 at various annual rates.		
22			
23	Annual	Number of	# years
24	Rate	Years	rounded up
25	1%	161.75	162
26	2%	81.27	82
27	3%	54.45	55
28	4%	41.04	42
29	5%	32.99	33
30	6%	27.62	28
31	7%	23.79	24
32	8%	20.91	21
33	9%	18.68	19
34	10%	16.89	17
35	11%	15.42	16
36	12%	14.20	15
37			
38			

Pack members did some quick checks on the spreadsheet output. For instance, the amount needed to accumulate to $25,000 after 10 years at 8% would be $25,000/(1.08)^{10} = $11,579.84. The number of years required for $5000 to grow to $25,000 at 8% would be $\log(5)/\log(1.08) = 20.91237$, basically 21 years. These and other checks verified the spreadsheet output as correct.

"Nicely done," said Herkimer.

In the meantime, the ever-curious Frances produced this spreadsheet relating to the financial material being discussed:

	A	B	C	D	E	F	G	H	I	J
1	Spreadsheet by FRANCES									
2	Single deposit required to accumulate to $100,000 after indicated number of years at various rates									
3			ANNUAL INTEREST RATES							
4			2%	4%	6%	8%	10%	12%		
5	Y	10	$82,035	$67,556	$55,839	$46,319	$38,554	$32,197		
6	E	20	$67,297	$45,639	$31,180	$21,455	$14,864	$10,367		
7	A	30	$55,207	$30,832	$17,411	$9,938	$5,731	$3,338		
8	R	40	$45,289	$20,829	$9,722	$4,603	$2,209	$1,075		
9	S	50	$37,153	$14,071	$5,429	$2,132	$852	$346		
10										
11										
12										
13										
14										
15										
16										
17										
18										
19										
20										
21										
22										
23										
24										
25										
26										
27										
28										
29										
30										
31										
32										
33										
34										
35										
36										
37										
38										

30

"Wow!" shouted Herkimer. "Frances' display clearly demonstrates the power of compounding. Look, if you had a single deposit of $346 that could earn 12% a year, you would have $100,000 after 50 years." Pack members checked this one out with calculators. Indeed, $100,000/(1.12)^{50} = $100,000(1.12)^{-50} = $346.02.

As he would do many times during his sessions with the Pack, Herkimer emphasized that many people simply don't appreciate the power of compounding. "If you start saving early in your life," he stated, "you can start off with very little and end up with a lot. The power of compounding is amazing."

GENERAL FORMULA FOR SINGLE DEPOSIT DISCOUNTING
(See Appendix A for algebraic derivation)
The single deposit required to accumulate to an amount A in n years at annual interest rate i is
$$A/(1 + i)^n = A(1+i)^{-n}$$
The total interest earned is $A - A/(1 + i)^n$

BASIC ALGEBRA: If $x \neq 0$, then $x^{-1} = 1/x$

HERKIMER'S FASCINATING FINANCIAL FACTS:

The U.S. Treasury estimates that a $1 bill lasts about 18 months, a $5 bill lasts two years, a $10 bill lasts three years, a $20 bill lasts for years, and $50 and $100 bills last nine years.

ACTIVITY SET FOR SESSION #4

1. **ALCAL** activity: Use your calculator to check the following amounts that appear on the spreadsheet constructed by Frances:

(a) The amount needed to accumulate to $100,000 after 30 years at rate 8%:

To be evaluated	Calculator entry	Number represented
$100,000(1.08)^{-30}$	100000*1.08^-30	

(b) The amount needed to accumulate to $100,000 after 50 years at rate 6%:

To be evaluated	Calculator entry	Number represented
$100,000(1.06)^{-50}$	100000*1.06^-50	

2. REFLECTION & COMMENT activity:

Those who don't understand interest pay it; those who do understand it earn it.

(Common financial proverb)

3. What single deposit made right now would accumulate to

(a) $10,000 in five years if the annual interest rate is 8%?

(b) $80,000 in ten years if the annual interest rate is 6.5%?

(c) $1,000,000 in twenty years if the annual interest rate is 5.8%?

4. At the indicated annual interest rate, how long would it take for

(a) $1000 to grow to $10,000 at 6%?

(b) $50,000 to increase of $200,000 at 4.9%?

(c) $3500 to grow to $15,000 at 5.3%?

5. Complete the following table:

Annual rate	Single deposit needed to accumulate to $1,000,000 in 40 years	Number of years required for $1,000 to accumulate to $1,000,000
1%		
2%		
3%		
4%		
5%		
6%		
7%		
8%		
9%		
10%		
11%		
12%		
13%		
14%		
15%		
16%		
17%		
18%		
19%		
20%		

$$$
$$

Session #5
TRAPPING A FINANCIAL SOLUTION

> *Remember that money is of a prolific, generating nature. Money can beget money, and its offspring can beget more, and so on. Five shillings turned is six; turned again it is seven; and so on until it becomes a hundred pounds. The more there is of it, the more it produces at every turning, so that the profits rise quicker and quicker.*
> -Benjamin Franklin

Herkimer placed this transparency on the overhead:

> **What interest rate would have an investment of $10,000 accumulate to $30,000 after 12 years?** If x is the desired rate, then x is a solution to the equation $10,000(1 + x)^{12} = 30000$.
>
> NOTE: $\$10,000(1.09)^{12} = \$28,127$ and $\$10,000(1.10)^{12} = \$31,384$.

"Can you tell me something about the requested rate?" asked Herkimer.

"Sure," replied Roger. "From the given information you can see that 9% is too little and 10% is too much. The rate is somewhere between 9% and 10%."

Herkimer praised Roger for a good observation. He then went on to explain that one could find a solution to the equation by a trial-and-error process, initially "trapping" the solution between 9% and 10%. Then one could test the values 9.1%, 9.2%, 9.3%, etc. and trap the solution in a smaller interval.

"Could someone find a solution algebraically?" inquired Herkimer.

Valarie went to the chalkboard. "This is second-year algebra stuff, but I can do it," she said as she began writing. "I'll be using fractional exponents. Remember them?"

$$10000(1+x)^{12} = 30000$$
$$\Rightarrow (1+x)^{12} = 3$$
$$\Rightarrow 1+x = 3^{1/12}$$
$$\Rightarrow x = 3^{1/12} - 1 = .0958727 \approx 9.587\%$$

ALGEBRA!
IF $x > 0$,
$x^{\frac{1}{N}} = \sqrt[N]{x}$.

"Well done," said Herkimer. He went on to explain that some financial equations are more complex and do not have an easily-obtainable algebraic solution. He then displayed

34

the output from a spreadsheet containing some handwritten material he had created to demonstrate the trapping of a solution to the equation $10,000(1 + x)^{12} = 30000$. The Pack noted that the "trapped" solution agreed with the one obtained by Valarie.

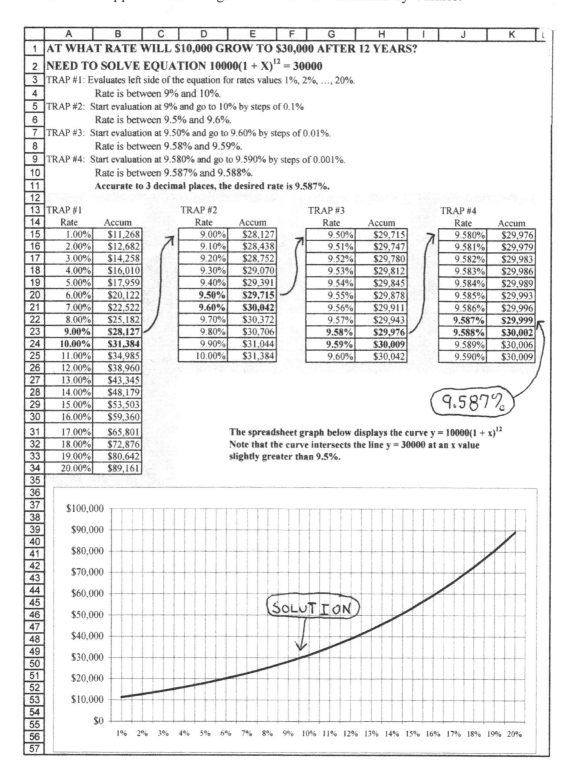

1	**AT WHAT RATE WILL $10,000 GROW TO $30,000 AFTER 12 YEARS?**	
2	**NEED TO SOLVE EQUATION $10000(1 + X)^{12} = 30000$**	
3	TRAP #1: Evaluates left side of the equation for rates values 1%, 2%, …, 20%.	
4	Rate is between 9% and 10%.	
5	TRAP #2: Start evaluation at 9% and go to 10% by steps of 0.1%	
6	Rate is between 9.5% and 9.6%.	
7	TRAP #3: Start evaluation at 9.50% and go to 9.60% by steps of 0.01%.	
8	Rate is between 9.58% and 9.59%.	
9	TRAP #4: Start evaluation at 9.580% and go to 9.590% by steps of 0.001%.	
10	Rate is between 9.587% and 9.588%.	
11	**Accurate to 3 decimal places, the desired rate is 9.587%.**	

	TRAP #1			TRAP #2			TRAP #3			TRAP #4	
14	Rate	Accum		Rate	Accum		Rate	Accum		Rate	Accum
15	1.00%	$11,268		9.00%	$28,127		9.50%	$29,715		9.580%	$29,976
16	2.00%	$12,682		9.10%	$28,438		9.51%	$29,747		9.581%	$29,979
17	3.00%	$14,258		9.20%	$28,752		9.52%	$29,780		9.582%	$29,983
18	4.00%	$16,010		9.30%	$29,070		9.53%	$29,812		9.583%	$29,986
19	5.00%	$17,959		9.40%	$29,391		9.54%	$29,845		9.584%	$29,989
20	6.00%	$20,122		**9.50%**	**$29,715**		9.55%	$29,878		9.585%	$29,993
21	7.00%	$22,522		**9.60%**	**$30,042**		9.56%	$29,911		9.586%	$29,996
22	8.00%	$25,182		9.70%	$30,372		9.57%	$29,943		**9.587%**	**$29,999**
23	**9.00%**	**$28,127**		9.80%	$30,706		**9.58%**	**$29,976**		**9.588%**	**$30,002**
24	**10.00%**	**$31,384**		9.90%	$31,044		**9.59%**	**$30,009**		9.589%	$30,006
25	11.00%	$34,985		10.00%	$31,384		9.60%	$30,042		9.590%	$30,009
26	12.00%	$38,960									
27	13.00%	$43,345									
28	14.00%	$48,179									
29	15.00%	$53,503									
30	16.00%	$59,360									
31	17.00%	$65,801									
32	18.00%	$72,876									
33	19.00%	$80,642									
34	20.00%	$89,161									

9.587%

The spreadsheet graph below displays the curve $y = 10000(1 + x)^{12}$
Note that the curve intersects the line $y = 30000$ at an x value slightly greater than 9.5%.

SOLUTION

Herkimer went on to explain that a graphics calculator could also be used to get solutions to equations by the trapping method.

As was true with all Pack members, Wayne was fascinated by the solution-trapping process. He wondered how long it would take $1,000 to grow to one million dollars at an annual interest rate of 20%. When Wayne produced a solution-trapping spreadsheet for the equation $1,000(1.20)^x = 1,000,000$, Herkimer asked him to display a hand-written algebraic solution on the sheet. Here is what Wayne produced:

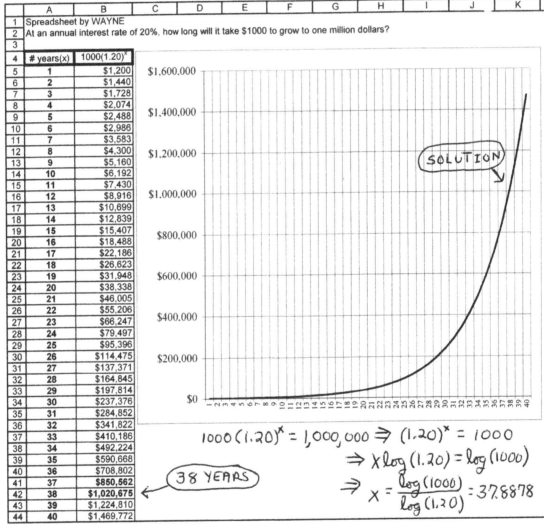

"Beautiful," said Herkimer. "We are now well set to move into the more sophisticated stages of financial mathematics."

HERKIMER'S FASCINATING FINANCIAL FACTS:

A woman's portrait has appeared on U.S. paper money. Martha Washington's portrait appeared on the face of $1 Silver Certificates in 1886 and 1891, and on the back of a $1 Silver Certificate of 1896.

ACTIVITY SET FOR SESSION #5

1. **ALCAL** activity: Use your calculator to check the following chalkboard presentation calculation made by Valarie:

To be evaluated	Calculator entry	Number represented
$3^{1/12} - 1$	3^(1/12)-1	

What do you get if you incorrectly do the calculator entry as 3^1/12-1? Why is this not a correct representation for what is desired?

2. REFLECTION & COMMENT activity:

Americans have caved into the emotional media hype, becoming so accustomed to spending and borrowing in order to answer consumerism's siren call that they never question whether something should be purchased. They only ask themselves if there will be enough money to make the minimum monthly payment. Even if there aren't enough funds to cover a monthly payment, many Americans buy a product anyway. We call this reckless spending the "disease of consumerism."

(*Money Mastery*, by Williams, Jeppson, and Botkin)

In activities 3-5 use the trial-and-error method to find a solution for the displayed equation. Then check the solution algebraically. [NOTE: These equations can be solved algebraically, but some financial equations do not have easy algebraic solutions.]

3. If we want to know the single deposit required that will accumulate to $50,000 after 20 years at an annual rate of 7.5%, we need a solution to the linear equation

$$x(1.075)^{20} = 50,000$$

4. If the annual interest rate is 8.4% and we want to know how many years will be required for $100 to accumulate to $500, we need a solution for the non-linear equation

$$100(1.084)^x = 500$$

5. If we wish to know the annual interest rate required for $100 to accumulate to $40,000 in 30 years, we need to solve the non-linear equation

$$100(1 + x)^{30} = 40,000$$

----

6. How many years will it take $2500 to accumulate to $50,000 if the annual interest rate is 5.4%?

7. If $2000 accumulates to $16,800 in ten years, what is the annual interest rate?

8. What single deposit will accumulate to $25,000 after eight years if the annual interest rate is 6.3%?

$$$
$$

HERKIMER'S QUICK QUIZ
SESSIONS 1-5.

Find the correct response for
each question in the 25 cells
at the bottom of the page.

1. With a *simple* annual interest rate of 9%, how much interest will a single deposit of
   $10,000 earn in ten years?

2. With a *compound* annual interest rate of 9%, how much interest will a single deposit of
   $10,000 earn in ten years?

3. What single deposit will accumulate to $30,000 in fifteen years at an annual interest rate
   of 6.7%?

4. How long would it take a single deposit to triple in value at an annual interest rate of
   11.3%?

5. What is the accumulated value of a single deposit of $6000 in thirty years if the annual
   interest rate is 5.4%?

6. A 20-year deposit of $4500 earns an annual rate of 6% for the first twelve years and an
   annual rate of 7.8% for the remaining eight years. What is the accumulation at the end
   of the twentieth year?

7. If $1000 accumulates to $2800 in 12 years, what is the annual interest rate earned?

8. If a single deposit of D dollars earns an annual rate of 5.9% for T years, write the
   algebraic expression representing the accumulation at the end the T years.

9. If a single deposit of D dollars accumulates to T dollars in 10 years at an annual interest
   rate of 5.9%, write an equation representing this financial statement.

10. If T dollars must be invested now at an annual rate of 5.9% to accumulate to D dollars
    in 10 years, write an equation representing this financial statement.

	A	B	C	D	E
1	8.96%	$D(1.059)^T$	8.14%	$22,345	11 years
2	$10,987	$22,854	$16,513	$D(1.059)^{-10} = T$	$T(1.59)^{10} = D$
3	$(1.059)^T = D$	9.97%	$17,833	$9000	7.45%
4	14.21%	19 years	$29,065	$T(1.059)^D$	$11,341
5	8 years	$D(1.059)^{10} = T$	14 years	11.23%	$13,674

Session #6
PACK MEMBERS CREATE USEFUL FINANCIAL TABLES

> *I've got all the money I'll ever need if I die by four o'clock this afternoon.*
> -Henny Youngman

Herkimer told the Pack that with modern calculators and computers, tables that assist in financial calculations are less necessary now than they were many years ago. With modern spreadsheets, the "old time" tables that appeared in financial textbooks can be easily constructed. This gave him an idea for a neat educational experience for the Pack.

He displayed tables from financial books that had been published many years ago. One such table displayed values of

$$(1 + i)^n$$

over a 20-year period for interest rate values 1%, 2%, 3%, ..., 9%, 10%. This table allowed one to compute the accumulation of $1 over many years at various interest rates. For instance, the table indicated that $(1.07)^{10} = 1.96715$. So, $1000 invested at 7% for 10 years would accumulate to $1000(1.96715) = $1967.15.

A second table showed values of

$$1/(1 + i)^n = (1 + i)^{-n}$$

over the same 20-year period and for the same interest rates. If, for instance, one wished to know the required deposit to accumulate to $1000 after 15 years at 7%, the table indicates that $(1.07)^{-15} = 0.36245$. Hence the required amount would be $1000(0.36245) = $362.45.

Herkimer asked the Pack to produce these tables on spreadsheets and to create four meaningful financial questions with solutions that could be answered easily using the tables. The group loved this type of challenge as they had become addicted to spreadsheet creation during their statistical studies with Herkimer.

This kept the Pack occupied for about twenty minutes. They came up with some nifty-looking tables and some excellent questions. Tables produced by the team of Brenda and Frances are displayed here.

	A	B	C	D	E	F	G	H	I	J	K	L
1	Spreadsheet by BRENDA											
2					VALUES OF $(1 + i)^n$							
3												
4			ANNUAL INTEREST RATES									
5			1%	2%	3%	4%	5%	6%	7%	8%	9%	10%
6	Y	1	1.01000	1.02000	1.03000	1.04000	1.05000	1.06000	1.07000	1.08000	1.09000	1.10000
7	E	2	1.02010	1.04040	1.06090	1.08160	1.10250	1.12360	1.14490	1.16640	1.18810	1.21000
8	A	3	1.03030	1.06121	1.09273	1.12486	1.15763	1.19102	1.22504	1.25971	1.29503	1.33100
9	R	4	1.04060	1.08243	1.12551	1.16986	1.21551	1.26248	1.31080	1.36049	1.41158	1.46410
10	S	5	1.05101	1.10408	1.15927	1.21665	1.27628	1.33823	1.40255	1.46933	1.53862	1.61051
11		6	1.06152	1.12616	1.19405	1.26532	1.34010	1.41852	1.50073	1.58687	1.67710	1.77156
12		7	1.07214	1.14869	1.22987	1.31593	1.40710	1.50363	1.60578	1.71382	1.82804	1.94872
13		8	1.08286	1.17166	1.26677	1.36857	1.47746	1.59385	1.71819	1.85093	1.99256	2.14359
14		9	1.09369	1.19509	1.30477	1.42331	1.55133	1.68948	1.83846	1.99900	2.17189	2.35795
15		10	1.10462	1.21899	1.34392	1.48024	1.62889	1.79085	1.96715	2.15892	2.36736	2.59374
16		11	1.11567	1.24337	1.38423	1.53945	1.71034	1.89830	2.10485	2.33164	2.58043	2.85312
17		12	1.12683	1.26824	1.42576	1.60103	1.79586	2.01220	2.25219	2.51817	2.81266	3.13843
18		13	1.13809	1.29361	1.46853	1.66507	1.88565	2.13293	2.40985	2.71962	3.06580	3.45227
19		14	1.14947	1.31948	1.51259	1.73168	1.97993	2.26090	2.57853	2.93719	3.34173	3.79750
20		15	1.16097	1.34587	1.55797	1.80094	2.07893	2.39656	2.75903	3.17217	3.64248	4.17725
21		16	1.17258	1.37279	1.60471	1.87298	2.18287	2.54035	2.95216	3.42594	3.97031	4.59497
22		17	1.18430	1.40024	1.65285	1.94790	2.29202	2.69277	3.15882	3.70002	4.32763	5.05447
23		18	1.19615	1.42825	1.70243	2.02582	2.40662	2.85434	3.37993	3.99602	4.71712	5.55992
24		19	1.20811	1.45681	1.75351	2.10685	2.52695	3.02560	3.61653	4.31570	5.14166	6.11591
25		20	1.22019	1.48595	1.80611	2.19112	2.65330	3.20714	3.86968	4.66096	5.60441	6.72750
26												
27												
28			*SOLUTIONS MUST BE OBTAINED USING TABLE. NO CALCULATORS ALLOWED.*									
29												
30												
31			QUESTIONS:									
32		(1)	To what amount will $1000 accumulate after 15 years at an annual rate of 6%?									
33		(2)	To what amount will $100,000 accumulate after 20 years at an annual rate of 8%?									
34		(3)	How long will it take a single deposit to double at an annual rate of 5%?									
35		(4)	How long will it take a deposit of $1 to grow to $5 at an annual rate of 10%?									
36												
37												
38												
39			ANSWERS (Numbers in **bold type** came from table):									
40		(1)	$1000(1.06)^{15}$ = $1000(**2.39656**) = $2396.56.									
41		(2)	$100,000(1.08)^{20}$= $100,000(**4.66096**) = $466,096.									
42		(3)	From the table, $(1.05)^{14}$ = **1.97993** and $(1.05)^{15}$ = **2.07893**. Answer: 15 years.									
43		(4)	From the table, $(1.10)^{16}$ = **4.59497** and $(1.10)^{17}$ = **5.05447**. Answer: 17 years.									

	A	B	C	D	E	F	G	H	I	J	K	L	M
1	Spreadsheet by FRANCES												
2													
3					VALUES OF $(1+i)^{-n}$								
4													
5			ANNUAL RATES										
6			1%	2%	3%	4%	5%	6%	7%	8%	9%	10%	
7	Y	1	0.99010	0.98039	0.97087	0.96154	0.95238	0.94340	0.93458	0.92593	0.91743	0.90909	
8	E	2	0.98030	0.96117	0.94260	0.92456	0.90703	0.89000	0.87344	0.85734	0.84168	0.82645	
9	A	3	0.97059	0.94232	0.91514	0.88900	0.86384	0.83962	0.81630	0.79383	0.77218	0.75131	
10	R	4	0.96098	0.92385	0.88849	0.85480	0.82270	0.79209	0.76290	0.73503	0.70843	0.68301	
11	S	5	0.95147	0.90573	0.86261	0.82193	0.78353	0.74726	0.71299	0.68058	0.64993	0.62092	
12		6	0.94205	0.88797	0.83748	0.79031	0.74622	0.70496	0.66634	0.63017	0.59627	0.56447	
13		7	0.93272	0.87056	0.81309	0.75992	0.71068	0.66506	0.62275	0.58349	0.54703	0.51316	
14		8	0.92348	0.85349	0.78941	0.73069	0.67684	0.62741	0.58201	0.54027	0.50187	0.46651	
15		9	0.91434	0.83676	0.76642	0.70259	0.64461	0.59190	0.54393	0.50025	0.46043	0.42410	
16		10	0.90529	0.82035	0.74409	0.67556	0.61391	0.55839	0.50835	0.46319	0.42241	0.38554	
17		11	0.89632	0.80426	0.72242	0.64958	0.58468	0.52679	0.47509	0.42888	0.38753	0.35049	
18		12	0.88745	0.78849	0.70138	0.62460	0.55684	0.49697	0.44401	0.39711	0.35553	0.31863	
19		13	0.87866	0.77303	0.68095	0.60057	0.53032	0.46884	0.41496	0.36770	0.32618	0.28966	
20		14	0.86996	0.75788	0.66112	0.57748	0.50507	0.44230	0.38782	0.34046	0.29925	0.26333	
21		15	0.86135	0.74301	0.64186	0.55526	0.48102	0.41727	0.36245	0.31524	0.27454	0.23939	
22		16	0.85282	0.72845	0.62317	0.53391	0.45811	0.39365	0.33873	0.29189	0.25187	0.21763	
23		17	0.84438	0.71416	0.60502	0.51337	0.43630	0.37136	0.31657	0.27027	0.23107	0.19784	
24		18	0.83602	0.70016	0.58739	0.49363	0.41552	0.35034	0.29586	0.25025	0.21199	0.17986	
25		19	0.82774	0.68643	0.57029	0.47464	0.39573	0.33051	0.27651	0.23171	0.19449	0.16351	
26		20	0.81954	0.67297	0.55368	0.45639	0.37689	0.31180	0.25842	0.21455	0.17843	0.14864	
27													
28													
29	*SOLUTIONS MUST BE OBTAINED USING TABLE. NO CALCULATORS ALLOWED.*												
30													
31													
32	QUESTIONS:												
33		(1)	What single deposit will accumulate to $10,000 in 12 years if the annual interest rate is 6%?										
34		(2)	If the annual interest rate is 10%, what deposit will accumulate to $100,000 in 20 years?										
35		(3)	If you wanted a deposit to double in 12 years, what annual interest rate would be required?										
36		(4)	If you wanted a deposit to triple in 15 years, what annual interest rate would be required?										
37													
38													
39													
40	ANSWERS (Numbers in **bold type** came from table):												
41		(1)	$10,000$(1.06)^{-12} = $10,000(**0.49697**) = $4969.70.										
42		(2)	$100,000$(1.10)^{-20} = $100,000(**0.14864**) = $14,864.										
43		(3)	Using the 12-year row, any rate that is 6% or more would result in the desired doubling.										
44			(Table indicates that it would require 56 cents to grow to $1 in 12 years at 5%.)										
45		(4)	Using the 15-year row, a rate somewhere between 7% and 8% would obtain the desired										
46			tripling. Any rate that is 8% or above would triple the deposit in 15 years.										
47			(Table indicates that it would require 36 cents to grow to $1 at 7% and that 32 cents would										
48			grow to $1 at 8%.)										

"Nifty pieces of spreadsheet work," concluded a very pleased Herkimer.

HERKIMER'S FASCINATING FINANCIAL FACTS:

While no portraits of African Americans have been on U.S. paper money, signatures of four African American men and one African American woman have appeared. The men (Blanche K. Bruce, Judson W. Lyons, William T. Vernon, and James C. Napier) served as Registers of the Treasury and the woman (Azie Taylor Morton) served as Treasurer of the United States.

ACTIVITY SET FOR SESSION #6

1. **ALCAL** activity: Use your calculator to check the following values in the tables presented by Brenda and Frances:

(a) In Brenda's table the value $(1.09)^{19}$:

To be evaluated	Calculator entry	Number represented
1.09^{19}	1.09^19	

(b) In Frances' table the value $(1.06)^{-14}$:

To be evaluated	Calculator entry	Number represented
1.06^{-14}	1.06^-14	

2. REFLECTION & COMMENT activity:

*Today's generation, instead of fearing that it will not have **anything**, fears that it will not have **everything**.*

(*Money Mastery*, by Williams, Jeppson, and Botkin)

3. Use Brenda's spreadsheet table to calculate

    (a) the accumulation of $100 for 18 years at an annual rate of 9%.

    (b) the accumulation of $100,000 for 13 years at an annual rate of 7%.

    (c) the number of years required for money to double at a 7% annual rate.

    (d) the number of years required for money to triple at an 8% annual rate.

4. Use Frances' spreadsheet table to calculate

    (a) the deposit needed to accumulate to $10,000 in 14 years at a 9% annual rate.

    (b) the deposit required to accumulate o $1,000,000 in 20 years at an annual rate of 8%.

5. If both spreadsheet tables were extended to include an 11% column

    (a) what would be the 20-year entry in Brenda's table?

    (b) what would be the 20-year entry in Frances' table?

    (c) Explain why the product of your answers to (a) and (b) should be 1.

6. Using Brenda's spreadsheet table, calculate

    (a) the interest earned on a single deposit of $10,000 earning an annual rate of 7% for 12 years.

    (b) the interest earned on a single deposit of $100,000 earning an annual rate of 8% for 20 years.

$$
$$

Session #7
THE CONCEPT OF FINANCIAL INFLATION

> *Part of the $10 million I spent on gambling, part on booze and part on women. The rest I spent foolishly.*
> -George Raft

Pack members were somewhat familiar with the concept of *inflation*. Herkimer thought it would be a good time to bring up this ever-present topic. He put this transparency on the overhead.

INFLATION: The rate at which the cost of goods and services is rising, and subsequently, purchase power is falling.

At a 4% inflation rate, an item selling for $1 now will cost $1(1 + .04) = $1(1.04) = $1.04 one year from now.

Two years from now that item would cost $1(1.04)^2 = $1.0816.

Three years from now that item would cost $1(1.04)^3 = $1.124864.

Ten years from now that item would cost $1(1.04)^{10} = $1.480244

Pack members were quick to notice that they had just constructed tables that would display these and similar cost increases at various rates over a period of years. "Indeed you did," said Herkimer, "and you even did a bit more than that." He went on to explain that the second table they constructed during the least session displayed how much a present dollar would be worth in future years at various inflation rates. For instance, if the inflation rate was a constant 4% for 6 years, then $1 today would be worth $1/(1.04)^6 = $0.7903145257 in 6 years. "In other words, today's dollar would just buy 79 cents worth of goods in 6 years."

Herkimer asked the group to produce some spreadsheet tables and graphics to illustrate the concept of inflation at various rates. He also asked them to put some examples below the tables to illustrate how they were to be used. The challenge was readily accepted by this group of intelligent and inquisitive students.

The team of Roger and Stephen came up with the following display:

	A	B	C	D	E	F	G	H	I	J	K	L	M	N	O
1	Spreadsheet by ROGER and STEPHEN														
2															
3	Inflation table: Displays cost of item presently valued at $1 after specified number														
4	of years at various inflation rates.														
5			INFLATION RATE												
6			1%	2%	3%	4%	5%	6%	7%	8%	9%	10%			
7	Y	1	$1.01	$1.02	$1.03	$1.04	$1.05	$1.06	$1.07	$1.08	$1.09	$1.10			
8	E	2	$1.02	$1.04	$1.06	$1.08	$1.10	$1.12	$1.14	$1.17	$1.19	$1.21			
9	A	3	$1.03	$1.06	$1.09	$1.12	$1.16	$1.19	$1.23	$1.26	$1.30	$1.33			
10	R	4	$1.04	$1.08	$1.13	$1.17	$1.22	$1.26	$1.31	$1.36	$1.41	$1.46			
11	S	5	$1.05	$1.10	$1.16	$1.22	$1.28	$1.34	$1.40	$1.47	$1.54	$1.61			
13		10	$1.10	$1.22	$1.34	$1.48	$1.63	$1.79	$1.97	$2.16	$2.37	$2.59			
14															
15	EXAMPLES:														
16	At a 4% inflation rate, an item costing $1 now will cost $1.22 in 5 years														
17	At a 6% inflation rate, an item costing $1 now will cost $1.79 in 10 years														
18															
19															
20															
21	This table displays the decreasing value of $1 at various inflation rates.														
22			INFLATION RATE												
23			1%	2%	3%	4%	5%	6%	7%	8%	9%	10%			
24	Y	1	$0.99	$0.98	$0.97	$0.96	$0.95	$0.94	$0.93	$0.93	$0.92	$0.91			
25	E	2	$0.98	$0.96	$0.94	$0.92	$0.91	$0.89	$0.87	$0.86	$0.84	$0.83			
26	A	3	$0.97	$0.94	$0.92	$0.89	$0.86	$0.84	$0.82	$0.79	$0.77	$0.75			
27	R	4	$0.96	$0.92	$0.89	$0.85	$0.82	$0.79	$0.76	$0.74	$0.71	$0.68			
28	S	5	$0.95	$0.91	$0.86	$0.82	$0.78	$0.75	$0.71	$0.68	$0.65	$0.62			
30		10	$0.91	$0.82	$0.74	$0.68	$0.61	$0.56	$0.51	$0.46	$0.42	$0.39			
31															
32	EXAMPLES:														
33	If inflation rate is 3%, then $1 now will be worth 86 cents in 5 years.														
34	If inflation rate is 9%, then $1 now will be worth 42 cents in 10 years.														
35															
36															
37	This chart illustrates the effects of a 4% inflation rate over a period of 20 years.														
38															
39		Yr	Cost												
40		1	$1.04												
41		2	$1.08												
42		3	$1.12												
43		4	$1.17												
44		5	$1.22												
45		6	$1.27												
46		7	$1.32												
47		8	$1.37												
48		9	$1.42												
49		10	$1.48												
50		11	$1.54												
51		12	$1.60												
52		13	$1.67												
53		14	$1.73												
54		15	$1.80												
55		16	$1.87												
56		17	$1.95												
57		18	$2.03												
58		19	$2.11												
59		20	$2.19												
60															
61	EXAMPLES:														
62	If inflation rate is 4%, prices will double in 18 years.														
63	If inflation rate is 4%, an item costing $1 now will cost $2.19 in 20 years.														

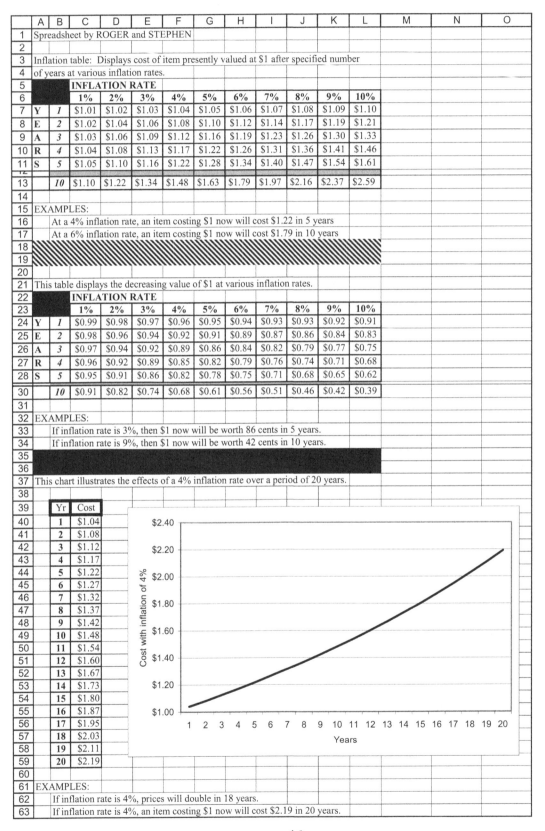

"Hey, hey, hey, a job well done," concluded Herkimer.

HERKIMER'S FASCINATING FINANCIAL FACTS:

The largest amount of coin money you can have without being able to make change for a dollar is $1.19. This involves having 3 quarters, 4 dimes, and 4 pennies.

ACTIVITY SET FOR SESSION #7

1. **ALCAL** activity: Use your calculator to check the following values in the table presented by Roger and Stephen:

(a) In the first section, the value of $(1.08)^{10}$:

To be evaluated	Calculator entry	Number represented
1.08^{10}	1.08^10	

(b) In the second section, the value $(1.09)^{-5}$:

To be evaluated	Calculator entry	Number represented
1.09^{-5}	1.09^-5	

2. REFLECTION & COMMENT activity:

Reducing inflation has costs in lost output and unemployment during the adjustment. Thus, an important question is whether zero percent inflation is sufficiently better for the economy that 2 to 3 percent inflation to warrant the effort of getting there.

(Rebecca Hellerstein, *Regional Review*, Winter 1997)

3. Complete the following inflation table indicating the cost of a $1 item after the indicated number of years at the specified inflation rates:

Inflation Rates

		12%	20%	25%	30%	50%
	1					
	5					
Years	10					
	20					
	25					

47

4. Complete the following inflation table to display the decreasing value of $1 after the indicated number of years at the specified inflation rates.

Inflation Rates

		12%	20%	25%	30%	50%
	1					
	5					
Years	10					
	20					
	25					

5. Here are financial terms and phrases related to the concept of inflation. Do a bit of research and provide a brief description for each item.

Consumer Price Index (CPI)	Cost-of-Living Index	Stagflation	Deflation
Hyperinflation	Federal Reserve	Demand-pull inflation	Cost-push inflation
Supply-side economics	Trickle-down economics	Gold standard	Money supply

$$$
$$$

48

Session #8
MULTIPLE DEPOSIT ACCUMULATION

> *It's good to have money and the things money can buy, but it's good, too, to check up once in a while and make sure you haven't lost the things money can't buy.*
> -George Horace Lorimer

Herkimer explained to the Pack that investments often involve a series of deposits or payments. He put this display on the overhead.

Suppose one makes a deposit of $1,000 at the beginning of each year for five years. If the deposits earn an annual interest rate of 8%, what is the value of this investment at the end of the fifth year?

Deposit	$1,000	$1,000	$1,000	$1,000	$1,000	A
End of Year	0	1	2	3	4	5

The Pack had already established five randomly-determined teams of two students each in preparation for a challenge they knew was coming. Herkimer put forth the challenge of calculating the value of A which he defined to be the accumulated value of the five payments after five years.

Roger and Valarie were first to respond. Roger began writing on the chalkboard while Valarie explained, "The first deposit earns interest for 5 years, the second for 4 years, the third for 3 years, the fourth for 2 years, and the fifth for 1 year." Roger completed his writing on the board and asked others to check the indicated total.

$$1000(1.08) + 1000(1.08)^2 + 1000(1.08)^3 + 1000(1.08)^4 + 1000(1.08)^5$$
$$= 1080.00 + 1166.40 + 1259.71 + 1360.49 + 1469.33$$
$$= 6335.93.$$

There was agreement that Roger's total was correct. Herkimer then asked the Pack to observe carefully the expression

$$\$1,000(1.08) + \$1,000(1.08)^2 + \$1,000(1.08)^3 + \$1,000(1.08)^4 + \$1,000(1.08)^5$$

Never one to overlook an opportunity to promote the important of basic algebra, he flipped this transparency on the overhead.

The Pack knew Herkimer was leading them down the road to an interesting result. "We can note that each of the five terms has a common factor of 1000(1.08)," he said. He then wrote on the chalkboard

$$A = \mathbf{\$1{,}000(1.08)}(1 + 1.08 + 1.08^2 + 1.08^3 + 1.08^3 + 1.08^4) = \$6{,}335.93.$$

Herkimer indicated that this computation is somewhat simpler on a calculator since $1,000 need only be typed in once, but it is still a bit cumbersome. After all, what would we be looking at if instead of 5 years we were talking about accumulating such payments for 20 years, or 30 years, or 50 years? For 50 years, it would be nice to be able to avoid summing 50 terms. There is a general formula that provides results for any number of years.

GENERAL FORMULA FOR MULTIPLE DEPOSIT ACCUMULATION
(See Appendix A for algebraic derivation)

						A
Deposit	D	D	D	------	D	D	.
Y = End of Year	0	1	2	------	n-2	n-1	n

The accumulated value A of a series of beginning-of-year deposits of D for n years at annual interest rate i is

$$A = D(1 + i)[(1 + i)^n - 1]/i$$

Herkimer then put up a spreadsheet relating to the computations of Roger and Valarie, and displaying use of the general formula he had just introduced.

	A	B	C	D	E	F
1	Deposit $1,000 at the beginning of each year for 5 years.					
2	Annual interest rate = 8%					
3						
4	Year	Deposit	Accumulation			
5	0	$1,000	$1,000.00			
6	1	$1,000	$2,080.00			
7	2	$1,000	$3,246.40			
8	3	$1,000	$4,506.11			
9	4	$1,000	$5,866.60			
10	5		$6,335.93			

←———— 1000 + 1.08(1000)

←———— 1000 + 1.08(2080)

←———— 1000 + 1.08(3246.40)

←———— 1000 + 1.08(4506.11)

←———— 1.08(5866.60)

$1000(1.08)\left[\dfrac{(1.08)^5 - 1}{.08}\right]$

Herkimer then provided examples of use of this formula on the overhead. He emphasized that the formula applied to a specific financial model, specifically beginning-of-year deposits for n years with no deposit at the end of the nth year.

Investment Description	Computation formula	Accumulation	Interest earned
Deposits of $5,000 at the beginning of each year for 10 years earning 7% annual rate	$5000(1.07)[(1.07)^{10} - 1]/.07$	**$73,918.00**	$73,918.00 - 10($5,000) = **$23,918.00**
Deposits of $5000 at the beginning of each year for 25 years earning 7% annual rate	$5000(1.07)[(1.07)^{25} - 1]/.07$	**$338,382.35**	$338,382.35 - 25($5,000) = **$213,382.35**
Deposits of $5000 at the beginning of each year for 20 years earning 5% annual rate.	$5000(1.05)[(1.05)^{20} - 1]/.05$	**$173,596.26**	$173,596.26 - 20($5,000) = **$73,596.26**
Deposits of $5000 at the beginning of each year for 20 years earning 10% annual rate.	$5000(1.10)[(1.10)^{20} - 1]/.10$	**$315,012.50**	$315,012.50 - 20($5,000) = **$215,012.50**

Carolyn noted something interesting. "Hey guys," she said, "notice that in the first example on the overhead you have invested a total of $50,000 and earned $23,918 in interest over the 10 year period." She went on to indicate that if you invested the entire $50,000 at the start of the period, you would have an accumulation of $50,000(1.07)^{10} = $98,357.57. The interest earned would be $48,357.57.

"Yup," responded Herkimer. "In the world of finance, $50,000 now is not the same as ten payments of $5,000 spread over a period of time. Carolyn has made an astute observation that is well worth noting."

In the work period following the discussion, Darren produced the following spreadsheet illustrating newly-learned financial concepts. Herkimer suggested that Carolyn's observation be noted on the sheet. Darren did this after his graphics display.

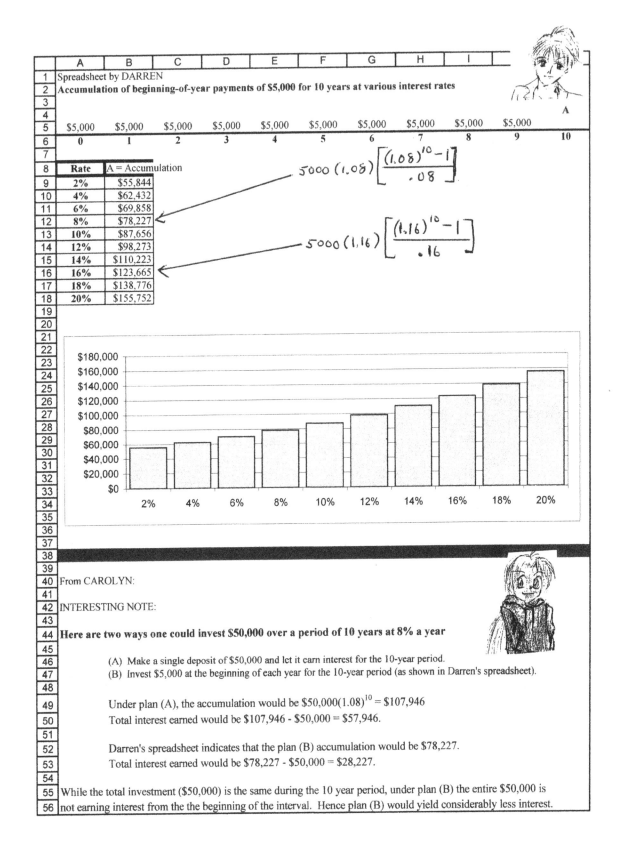

	A	B	C	D	E	F	G	H	I	
1	Spreadsheet by DARREN									
2	**Accumulation of beginning-of-year payments of $5,000 for 10 years at various interest rates**									
3										
4										
5	$5,000	$5,000	$5,000	$5,000	$5,000	$5,000	$5,000	$5,000	$5,000	$5,000
6	0	1	2	3	4	5	6	7	8	9	10
7										
8	**Rate**	A = Accumulation								
9	2%	$55,844								
10	4%	$62,432								
11	6%	$69,858								
12	8%	$78,227								
13	10%	$87,656								
14	12%	$98,273								
15	14%	$110,223								
16	16%	$123,665								
17	18%	$138,776								
18	20%	$155,752								

$$5000(1.08)\left[\frac{(1.08)^{10}-1}{.08}\right]$$

$$5000(1.16)\left[\frac{(1.16)^{10}-1}{.16}\right]$$

40	From CAROLYN:
42	INTERESTING NOTE:
44	**Here are two ways one could invest $50,000 over a period of 10 years at 8% a year**
46	(A) Make a single deposit of $50,000 and let it earn interest for the 10-year period.
47	(B) Invest $5,000 at the beginning of each year for the 10-year period (as shown in Darren's spreadsheet).
49	Under plan (A), the accumulation would be $50,000(1.08)^{10}$ = $107,946
50	Total interest earned would be $107,946 - $50,000 = $57,946.
52	Darren's spreadsheet indicates that the plan (B) accumulation would be $78,227.
53	Total interest earned would be $78,227 - $50,000 = $28,227.
55	While the total investment ($50,000) is the same during the 10 year period, under plan (B) the entire $50,000 is
56	not earning interest from the the beginning of the interval. Hence plan (B) would yield considerably less interest.

52

ACTIVITY SET FOR SESSION #8

1. **ALCAL** activity: Use your calculator to check the following values that appear in the table on Darren's spreadsheet:

   (a) Accumulation of beginning-of year deposits of $5,000 for 10 years at 6%:

To be evaluated	Calculator entry	Number represented
$5000(1.06)[(1.06)^{10}-1]/.06$	5000*1.06*(1.06^10-1)/.06	

   (b) Accumulation of beginning-of-year deposits of $5,000 for 10 years at 12%:

To be evaluated	Calculator entry	Number represented
$5000(1.12)[(1.12)^{10}-1]/.12$	5000*1.12*(1.12^10-1)/.12	

2. REFLECTION & COMMENT activity:

   You can't spend your way to happiness. But lots of folks try, setting themselves up for a lifetime of hefty credit card bills and emotional disappointment. You know the cycle: You see something in the store, you decide you just have to have it, you pony up the bucks and, a few weeks later, the purchase is all but forgotten - and you're hankering after something else. Academics refer to this as the hedonic treadmill. The lesson? If you want happiness, you won't find it at the shopping mall.

   (James Clements, personal finance writer, *Wall Street Journal*, November 19, 2006)

3. Find the accumulation for these beginning-of-year deposits.

   (a) $25,000 for 15 years at 9%.

   (b) $10,000 for 30 years at 7.2%.

   (c) $1,000 for 50 years at 8.6%.

53

4. (a) A single deposit of $20,000 earns interest at an annual rate of 7.35%. Find the accumulated value of this investment and the interest earned after 20 years.

   (b) Twenty beginning-of-year deposits of $1,000 earn an annual rate of 7.35%. Find the accumulated value of this investment and the interest earned after 20 years.

   (c) In both (a) and (b), a total of $20,000 was invested at 7.35%. Explain the difference in the answers for (a) and (b).

5. An investment consists of beginning-of-year deposits of $4,000 for 10 years.

   (a) Find the accumulated value of the deposits at the end of the 10th year if the annual interest rate is 6%.

   (b) Find the accumulated value of the deposits at the end of the 10th year if the annual interest rate is 12%.

   (c) Is the interest earned in (b) exactly twice the interest earned in (a)?

6. Investment X consists of beginning-of-year deposits of $1,000 for ten years earning an annual rate of 10%. Investment Y consists of beginning-of-year deposits of 2,000 for ten years earning an annual rate of 5%. At the end of the 10th year, which investment

   (a) has the greatest accumulation?

   (b) has earned the most interest?

$$\$$$
$$\$$$

Session #9
FINANCIAL SITUATIONS INVOLVING PAYMENT ACCUMULATIONS

> *All I ask is the chance to prove that money can't make me happy.*
> -Spike Milligan

"OK," said Herkimer, "we're moving on into really *sophisticated* financial settings. Note that I didn't say *difficult* financial settings. It's simply a case of applying what you have learned. That's how life works in general."

"We're ready," responded Darren. "Sock it to us."

Herkimer put a transparency containing two problems on the overhead.

1.

What equal beginning-of-year deposits would be required to accumulate to $10,000 at the end of five years if the annual interest rate is 8%?

						$10,000
Deposit	x	x	x	x	x	.
End of Year	0	1	2	3	4	5

****

2.

If five beginning-of-year deposits of $1400 accumulate to $10,000 at the end of five years, what is the annual interest rate?

						$10,000
Deposit	$1400	$1400	$1400	$1400	$1400	.
End of Year	0	1	2	3	4	5

Frances was quick to respond to the first problem. "This involves nothing other than a linear equation," she said as she began writing on the chalkboard.

$$x(1.08)\left[\frac{(1.08)^5 - 1}{.08}\right] = \$10,000$$

$$\Rightarrow x(6.335929037) = \$10,000$$

$$\Rightarrow x = \frac{\$10,000}{6.335929037} = \$1,578.30.$$

Other Pack members confirmed that Frances' computation was correct. In the meantime, Stephen wrote an equation and other observations relating to obtaining a solution for problem #2.

$$\$1,400(1+i)\left[\frac{(1+i)^5-1}{i}\right] = \$10,000.$$

we need to find the value of i that makes left member of equation equal to $10,000.

The Pack knew there was no easy algebraic method to solve the equation. They realized that the trial-and-error trapping method they had previously experienced would produce solution. Herkimer asked the group to use their calculators to obtain a solution accurate to two decimal places.

Identifying the left member of the equation as A, Carolyn tested $i = 10\%$ and this produced a value A = 9401.85. "This is too small," she said. "The value of i we want is greater than 10%."

Wayne found that A = 9961.26 if $i = 12\%$ and A = 10,251.79 if $i = 13\%$. "The rate is between 12% and 13%," he claimed.

Working from Wayne's conclusion, Valarie discovered that if $i = 12.1\%$ then A = 9989.99 and that if $i = 12.2\%$ then A = 10,018.76. This trapped the rate between 12.1% and 12.2%.

 Finally, Roger computed that A = 9998.62 if $i = 12.13\%$ and that A = 10,001.50 if $i = 12.14\%$. Of these two values of A, the number 9998.62 was closest to 10,000. It was concluded that the desired solution was 12.13%. In other words, if 5 beginning-of-year deposits of $1400 accumulate to $10,000 after five years, the investment earns an annual rate of 12.13%.

"OK, good job," concluded Herkimer. "As is true in any branch of mathematics, if you can set up the equation to be solved and if there is a solution, you can find it by trial-and-error and the trapping process."

56

ACTIVITY SET FOR SESSION #9

Deposit	x	x	x	...	x	x	**A**
End of Year	0	1	2	...	n-2	n-1	n

x = beginning-of-year deposits.
n = number of years.
i = annual interest rate.
A = accumulated value of deposits at end of n-th year.

The financial formula relating these four variables is

$$A = x(1+i)[(1+i)^n - 1]/i$$

1. **ALCAL** activity: Use your calculator to check Frances' chalkboard computation
   relating to the deposits required to accumulate to $10,000 in five years at an 8% interest
   rate:

To be evaluated	Calculator entry	Number represented
$10000(.08)/[1.08(1.08^5-1)]$	10000*.08/(1.08*(1.08^5-1))	

2. REFLECTION & COMMENT activity:

   *In the new economic order, slow spenders - people who wait to buy and pay as they go - will
   be the ones to envy. They won't have the latest iPhone or the most stuff. But they'll have the
   best credit, actual savings, and stable finances - the new gateways to financial freedom.
   Slow spenders will be the ones able to get a loan to snap up property at a bargain. They'll
   have a rainy-day fund and some spare cash for their kids' college tuition. The microwave
   might still pop the popcorn, but gratification will be a lot less instant elsewhere and it might
   even taste better that way.*

   (Rick Newman, *Slow Down on Spending,* U.S. News & World Report, Dec, 2008)

In problems 3-6, use the formula above activity #1 to produce the requested information.

3. What equal beginning-of-year deposits are needed to accumulate to $50,000 at the end
   of 25 years if the annual interest rate is 6%?

4. What equal beginning-of-year deposits are needed to accumulate to one million dollars end of 40 years if the annual interest rate is 7.2%?

5. If beginning-of-year deposits of $3000 accumulate to $190,000 after 20 years, what is the annual interest rate for this investment? (Set up equation to be solved and use trial-and-error process.)

6. If beginning-of-year deposits of $1000 accumulate to $220,000 after 50 years, what is the annual interest rate for this investment? (Set up equation to be solved and use trial-and-error process.)

$$
$$

Session #10
OTHER MULTIPLE DEPOSIT ACCUMULATION MODELS

> *The easiest way for your children to learn about money is for you to not have any.*
> -Katharine Whitehorn

It was a new day. Using the overhead, Herkimer reminded the Pack that in a previous session they had worked with this financial model.

							A
Deposit	D	D	D	------	D	D	
Y = End of Year	0	1	2	------	n-2	n-1	n

The accumulated value **A** of a series of beginning-of-year deposits of D dollars for n years at annual interest rate i is

$$A = D(1 + i)[(1 + i)^n - 1]/i$$

"There are other models," he said. "If the payments are delayed for a year, the model looks like this."

(See Appendix A for algebraic derivation)

							A
Deposit		D	D	------	D	D	D
Y = End of Year	0	1	2	------	n-2	n-1	n

The accumulated value **A** of a series of end-of-year deposits of D dollars for n years at annual interest rate i is

$$A = D[(1 + i)^n - 1]/i$$

 The ever-alert Janice responded, "Hey, that second formula make sense. That formula is like the first one but simply missing a factor of $(1 + i)$. Since each deposit is earning interest for one year less, the second accumulation formula should be missing that factor."

"Wonderful observation," replied Herkimer. "Kudos to Janice for recognizing the power of basic algebra. Now let's look at one more common model. Note that in this situation we have a simple extension of the beginning-of-year model with one extra deposit that earns no interest."

							A
Deposit	D	D	D	------ D	D	D.	
Y = End of Year	0	1	2	------ n-2	n-1	n	

The accumulated value **A** of this series of payments for n years at annual interest rate i is

$$A = D + D(1 + i)[(1 + i)^n - 1]/i$$

Herkimer then put up a transparency illustrating the three models. Holding the transparency pen in his hand, he asked for volunteers to display the computations involved to find the respective accumulations at Y = 6 using formulas and an annual interest rate of 7%. Brenda, Wayne and Valarie took on the challenges.

								A
D	$500	$500	$500	$500	$500	$500		.
Y	0	1	2	3	4	5	6	

$$A = \$500(1.07)\left[\frac{(1.07)^6 - 1}{.07}\right] = \$3,827.01.$$

$$\text{INTEREST} = \$3,827.01 - 6(\$500) = \$827.01.$$

								A
D		$500	$500	$500	$500	$500	$500	
Y	0	1	2	3	4	5	6	

$$A = \$500\left[\frac{(1.07)^6 - 1}{.07}\right] = \$3,576.65.$$

$$\text{INTEREST} = \$3,576.65 - 6(\$500) = \$576.65.$$

								A
D	$500	$500	$500	$500	$500	$500	$500	
Y	0	1	2	3	4	5	6	

$$A = \$500 + \$500(1.07)\left[\frac{(1.07)^6 - 1}{.07}\right]$$
$$= \$500 + \$3,827.01$$
$$= \$4,327.01. \quad \text{Interest} = \$827.01.$$

ACTIVITY SET FOR SESSION #10

1. **ALCAL** activity: Use your calculator to check Brenda's computation relating to the accumulation of six beginning-of-year deposits of $500 for six years at 7%:

To be evaluated	Calculator entry	Number represented
$500(1.07)(1.07^6-1)/.07$	500*1.07*(1.07^6-1)/.07	

2. REFLECTION & COMMENT activity:

So many people believe we operate with a fixed-pie economy. They think money is gone forever if it leaves one place, but this is a faulty premise because it can just as easily go to another place. If you slice a (birthday) cake and give yourself a bigger piece, there's less for me. But money isn't like the cake, it's like the candles! If you light a candle and use it to light other candles, no candle is diminished. There is even more light.

(Dave Ramsay, author of *The Total Money Makeover*)

In activities 3 - 7 find the accumulated value and interest earned for the displayed deposits at the last year displayed.

3. Annual interest rate = 11%

Deposit		$800	$800	------	$800	$800	$800.
Y = End of Year	0	1	2	------	48	49	50

4. Annual interest rate = 7.6%.

Deposit		$10,000	$10,000	$10,000	------	$10,000	$10,000	$10,000.
Y = End of Year	0	1	2	------	23	24	25

61

5. Annual interest rate = 6.5%.

Deposit		$1000	$1000	------	$1000	$1000	$1000.
Y = End of Year	0	1	2	------	18	19	20

6. Annual interest rate = 5.8%

Deposit		$600	$600	$600	------	$600	$600	.
Y = End of Year	0	1	2	------	38	39	40

7. Annual interest rate = 12.7%

Deposit		$5000	$5000	$5000	------	$5000	$5000	$5000 .
Y = End of Year	0	1	2	------	13	14	15

$$
$$$

HERKIMER'S QUICK QUIZ
SESSIONS 6-10.

Find the correct response for
each question in the 25 cells
at the bottom of the page.

1. If the inflation rate is 3.2% an item presently selling for $20 will cost _____ in eight years.

2. At a 2.9% inflation rate, an item presently valued at $100 will have a value of _____ in ten years.

3. The accumulation of beginning-of-year deposits of $2000 for fifteen years will be _____ at the end of the fifteenth year if the annual interest rate is 5.25%.

4. What equal beginning-of-year deposits will accumulate to $60,000 after 15 years if the annual interest rate is 6.1%?

5. If beginning-of-year deposits of $5000 accumulate to $56,000 after eight years, what is the annual interest rate earned on this investment?

6. What is the solution to the equation $300(1+x)^{20} = \$1120$?

7. What is the solution to the equation $100(1+x)^2 + 100(1+x) = \$250$?

Questions 8-10 assume an annual interest rate of 5.65%. In each case, what is the accumulated value A at the end of the fourth year?

8.
					A
Deposit	$900	$900	$900	$900	$900
Year	0	1	2	3	4

9.
					A
Deposit	$900	$900	$900	$900	.
Year	0	1	2	3	4

10.
					A
Deposit		$900	$900	$900	$900
Year	0	1	2	3	4

	A	B	C	D	E
1	$4284	$3917	9.23%	13.37%	$1987
2	$46, 287	$20.44	12.50%	$5294	7.44%
3	6.25%	$48,766	$29.44	$75.14	15.83%
4	$4138	$5038	8.84%	$24.57	$51,365
5	$84.36	$2411	$4565	6.81%	$25.73

Session #11
SOME DEPOSITS DON'T FIT A NICE ACCUMULATION MODEL

> *Whoever said money can't buy happiness simply didn't know where to go shopping.*
> -Bo Derek

Herkimer definitely wanted to make the point that many times deposits do not represent a convenient series of equal payments. In many of these instances, accumulations must be done one deposit at a time. He put this example on the projector screen:

											A
D	$1000		$5000					$8000			.
Y	0	1	2	3	4	5	6	7	8	9	10

If deposits earn an annual rate of 7.5%, then the accumulation at the end of ten years is

$$A = \$1,000(1.075)^{10} + \$5,000(1.075)^7 + \$8,000(1.075)^2 = \$19,601.28.$$

Interest earned = $19,601.28 - $14,000.00 = $5,601.28.

Pack members had no problem with this thought. They also realized that modern technology makes such computations relatively easy.

"OK," said Herkimer, "suppose we make the following business arrangement. You lend me $10,000 now, $5000 one year from now, and $12,000 eight years from now. In return, I will pay you $50,000 ten years from now. You are lending me a total of $27,000 and you are getting $50,000 at a future date. What annual interest rate is your investment earning?"

"I'm with it," replied Darren. He went on to say that the answer to the question would require a solution to a non-linear equation. It didn't take the Pack long to produce the equation.

$$10,000(1 + i)^{10} + 5000(1 + i)^9 + 12,000(1 + i)^2 = 50,000$$

They realized the trail-and-error solution-trapping method they recently learned could be well-applied here.

"Right on!" responded Herkimer. "Now let's get teams working on a neat spreadsheet solution to this. And throw in some nice graphic displays in the process."

The team of Janice and Glen produced the following spreadsheet after using their calculators to convince themselves that the value of i was between 9% and 10%. At 9%, the investment value was $48,790 after ten years.   At 10% it was $52,247.

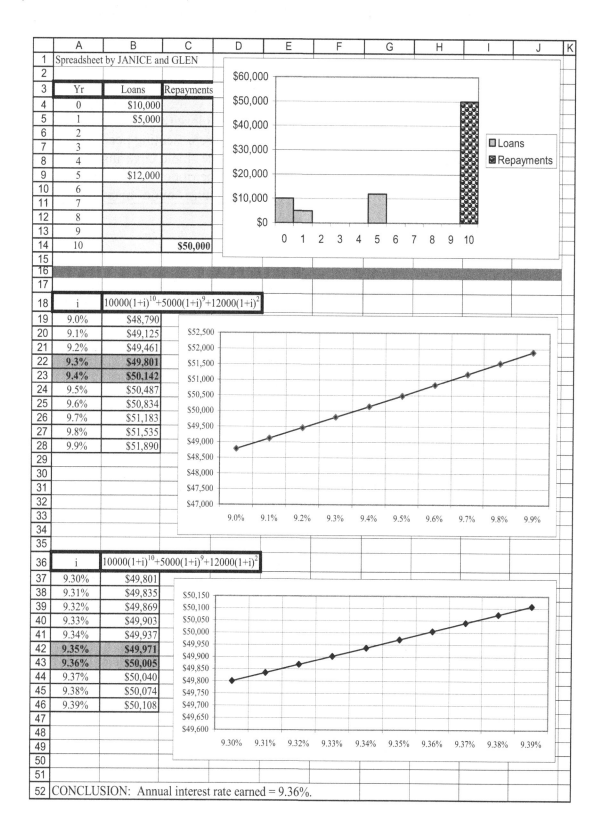

	A	B	C	D	E	F	G	H	I	J	K
1	Spreadsheet by JANICE and GLEN										
2											
3	Yr	Loans	Repayments								
4	0	$10,000									
5	1	$5,000									
6	2										
7	3										
8	4										
9	5	$12,000									
10	6										
11	7										
12	8										
13	9										
14	10		$50,000								
15											
16											
17											
18	i	$10000(1+i)^{10}+5000(1+i)^{9}+12000(1+i)^{2}$									
19	9.0%	$48,790									
20	9.1%	$49,125									
21	9.2%	$49,461									
22	9.3%	$49,801									
23	9.4%	$50,142									
24	9.5%	$50,487									
25	9.6%	$50,834									
26	9.7%	$51,183									
27	9.8%	$51,535									
28	9.9%	$51,890									
29											
30											
31											
32											
33											
34											
35											
36	i	$10000(1+i)^{10}+5000(1+i)^{9}+12000(1+i)^{2}$									
37	9.30%	$49,801									
38	9.31%	$49,835									
39	9.32%	$49,869									
40	9.33%	$49,903									
41	9.34%	$49,937									
42	9.35%	$49,971									
43	9.36%	$50,005									
44	9.37%	$50,040									
45	9.38%	$50,074									
46	9.39%	$50,108									
47											
48											
49											
50											
51											
52	CONCLUSION: Annual interest rate earned = 9.36%.										

65

"Beautifully done," concluded Herkimer. "I couldn't have done a better job myself. You young folks are really developing a knack for this investment material. I'm proud of you."

HERKIMER'S FASCINATING FINANCIAL FACTS:

On the back of a $1 bill above the American bald eagle is a glory (burst of light) with 13 stars. The right claw holds an olive branch with 13 leaves (representing peace) and the left claw holds a bundle of 13 arrows (symbolizing war). The shield covering the eagle's breast represents a united nation and contains 13 stripes. The ribbon held in the eagle's beak bears the 13-letter Latin motto *E Pluribus Unum*, meaning "out of many, one."

ACTIVITY SET FOR SESSION #11

1. **ALCAL** activity: Use your calculator to check the computation Herkimer displayed on the overhead projector at the beginning of the session:

To be evaluated	$1000(1.075)^{10}+5000(1.075)^{7}+8000(1.075)^{2}$
Calculator entry	1000*1.075^10+5000*1.075^7+8000*1.075^2
Number represented	

2. REFLECTION & COMMENT activity:

> *Not teaching your kids about money is like not caring whether they eat. If they enter the world without financial knowledge, they will have a much harder go of it. Make sure you let them in on your way of thinking about money - how you manage expenses, how you save, where you invest.*
>
> (Donald Trump, *How to Get Rich*)

3. For each set of displayed payments, find the investment value at the indicated annual rate at the end of the last year displayed on the line graph.

(a) Annual interest rate = 7%.

D	$500		$700	$900	$200				$600		.
Y	0	1	2	3	4	5	6	7	8	9	10	11	12

(b) Annual interest rate = 5.85%.

D	$10,000	$20,000		$50,000		$60,000		.
Y	0	1	2	3	4	5	6	7	8

(c) Annual interest rate = 6.7%.

D	$50,000	$40,000	$30,000			.
Y	0	1	2	3	4	5

4. In the two diagrams below, the indicated deposits accumulate to the amount **A** displayed on the line graph. Find the annual interest rate earned accurate to 2 decimal places. (ie. 7.64%). Clearly indicate the equation that must be solved to find the requested rate.

(a)

A = $30,000

D	$600			$700	$200		$800					
Y	0	1	2	3	4	5	6	7	8	9	10	11	12

(b)

A = $4000

D	$900	$300		$400			$700		$200			
Y	0	1	2	3	4	5	6	7	8	9

$$
$$

Session #12
PRESENT VALUE OF FUTURE PAYMENTS

> *I made my money the old fashioned way. I was very nice to a wealthy relative right before he died.*
> -Malcolm Forbes

Herkimer started the session with a display of excitement. "OK," he bellowed, "we are almost to the place where we can talk intelligently about loans, mortgages, and credit cards. We need only to talk about discounting a series of payments." He put this transparency on the overhead and asked the Pack to relate this to something to which they were exposed a few sessions earlier.

Suppose one wanted to withdraw $1,000 at the end of each year for the next 5 years. If a deposit can earn an annual rate of 7%, what amount **P** must be invested now to allow for the desired withdrawals?

	P					
Deposit		$1,000	$1,000	$1,000	$1,000	$1,000.
End of Year	0	1	2	3	4	5

P is the *present value* of the displayed payments at a 7% annual rate.

After a brief period of time, Frances realized this was nothing other than the sum of a series of single payment discounts. She wrote the following on the chalkboard:

$$P = \frac{\$1000}{1.07} + \frac{\$1000}{(1.07)^2} + \frac{\$1000}{(1.07)^3} + \frac{\$1000}{(1.07)^4} + \frac{\$1000}{(1.07)^5}$$

$$= \$4,100.20.$$

"You've got it," said Herkimer. "And, do you realize that you are looking at another geometric series. That good old algebra is going to come in handy once again."

"My gosh," chimed in Wayne, "do you mean there is another handy financial formula for present value payments?"

Herkimer responded: "For a series of equal payments like the one in this example, there is a nice financial formula. Go get it."

Using math power, the Pack met the challenge.

GENERAL FORMULA FOR PRESENT VALUE OF FUTURE PAYMENTS
(See Appendix A for algebraic derivation)

		P						
Deposit			D	D	------	D	D	D .
Y = End of Year	0		1	2	------	n-2	n-1	n

If the annual interest rate is i the single deposit **P** (present value) required to withdraw the displayed year-end payments D is

$$P = D[1-(1+i)^{-n}]/i.$$

Pack members used the formula to check Frances' chalkboard computations. Sure enough, $\$1000[1 - (1.07)^{-5}]/.07 = \$4,100.20$. Brenda and Janice produced a spreadsheet illustrating the new concept of present value.

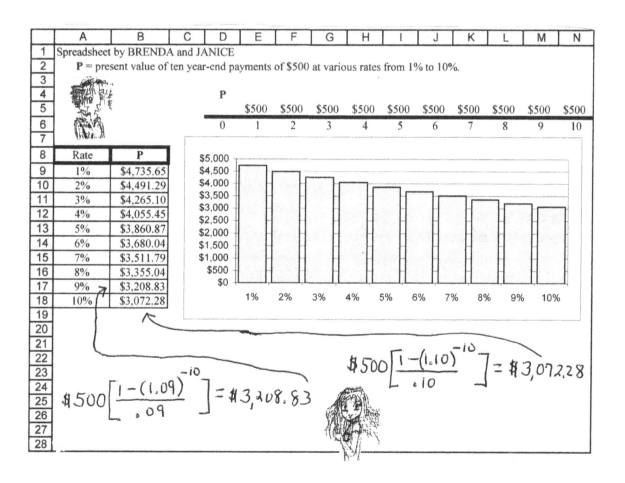

"Nicely done," concluded Herkimer. "We're making great progress towards financial literacy. Your future children will have parents who know how to manage money."

Prior to moving on, Herkimer wanted to emphasize that the present value formula was simply a linear equation in terms of the variables P and D. "Suppose," he said, "I deposited $25,000 now earning an annual interest rate of 7% and wanted to make annual equal year-end withdrawals for 15 years. What equation would I need to solve?"

Pack members quickly realized that the required equation is

$$\$25,000 = D[1 - (1.07)^{-15}]/.07$$

Calculators became active. It was agreed that the solution to this **linear** equation is D = $2744.87. The total interest earned would be 15($2744.87) - $25,000 = $16,173.05.

HERKIMER'S FASCINATING FINANCIAL FACTS:

In 1782 the US decided that its basic unit of currency would be the Spanish dollar or peso. The symbol for the peso was $, supposedly an ancient Phoenician sign indicating strength and sovereignty. Contrary to an existing belief, the symbol $ is not derived from the initials U and S superimposed on each other.

ACTIVITY SET FOR SESSION #12

1. **ALCAL** activity: Use your calculator to check the following values that appear in the spreadsheet table produced by Brenda and Janice:

(a) Present value of 10-year end payments of $500 at 6%:

To be evaluated	Calculator entry	Number represented
$500[1-(1.06)^{-10}]/.06$	500*(1-1.06^10)/.06	

(b) Present value of 10-year end payments of $500 at 8%:

To be evaluated	Calculator entry	Number represented
$500[1-(1.08)^{-10}]/.08$	500*(1-1.08^10)/.08	

2. REFLECTION & COMMENT activity:

> *An inflationary attack on top of everything else would steadily erode how much each of our remaining dollars would buy. The unemployed, those living on fixed incomes and the poor, would be hit the hardest, just as they were in the 1970's.*

(Mark Davis, McClatchy News Service, June 2009)

3. If you wanted to withdraw $10,000 at the end of each year for the next 20 years, what single deposit earning an annual rate of 7.2% would be required at the present time? What is the total amount of interest earned?

4. What single deposit earning an annual rate of 4.9% would be required now in order to withdraw year-end payments of $5,000 for the next 30 years? What is the total amount of interest earned?

5. If I deposit $20,000 now at an annual rate of 8%, what equal withdrawals can I make at the end of each year for

   (a) the next 10 years?   (b) the next 20 years?

   (c) the next 30 years?   (d) the next 50 years?

6. A deposit can earn an annual rate of 8.8%. What single deposit must I make now in order to withdraw $1000 one year from now, $2000 two years from now, $5000 three years from now, and $10,000 four years from now?

$$\$$$
$$\$$$

Session #13
MORE ON CALCULATING INVESTMENT YIELDS

> *A bargain is something you can't use at a price you can't resist.*
> -Franklin Jones

Herkimer continued to emphasize to the Pack that the most important thing in equation solving in real life situations is to create the equation that has to be solved. He put the following transparency on the overhead and noted that the phrase *annual yield* is often used in place of annual interest rate when investments are discussed. And, someone who makes a loan often does it as an investment. The lender usually gets back a total that is greater than the original loan value.

Loan #1:

You lend me $3500 now. I will pay you back with year-end payments of $1000 for the next five years. What is the annual yield on your investment?

	$3500					
Deposit		$1000	$1000	$1000	$1000	$1000.
End of Year	0	1	2	3	4	5

Equation to be solved: $\$3500 = \$1000[1 - (1 + i)^{-5}]/i$

****

Loan #2:

You lend me $5000 now. I will pay you back over the next 3 years with the following year-end payments. What is the annual yield on this investment?

	$4000			
Deposit		$1000	$2000	$3000
End of Year	0	1	2	3

Equation to be solved: $\$4000 = \$1000/(1+i) + \$2000/(1+i)^2 + \$3000/(1+i)^3$

The Pack knew the equations were non-linear and hence not trivial. Carolyn took her calculator and used the *intersect* feature that produced the point of intersection of two graphs to solve the loan #1 problem. Using x for i and noting that exponents are inserted with the ^ symbol, she defined one function to be Y1 = 1000*(1-(1+x)^-5)/x and a second function Y2 = 3500. Using the *intersect* feature, she found that the value of x was 0.13201588, or about 13.20%. Other Pack members used the trail-and-error trapping method and came to the same conclusion.

Thinking in terms of functions for Loan #2, Valarie defined a function P to be

$$P(i) = \$1000(1+i)^{-1} + \$2000(1+i)^{-2} + \$3000(1+i)^{-3}$$

She found that P(0.19) = \$4032.91 and P(0.20) = \$3958.33. Hence the requested annual yield was between 19% and 20%. Further trapping produced a rate of 19.44%. Other Pack members agreed with this result.

Darren and Janice produced a spreadsheet displaying a graphics approach to the questions relating to the two loans. Using pencil markings, they showed that the charts could be used to obtain approximate yield rates.

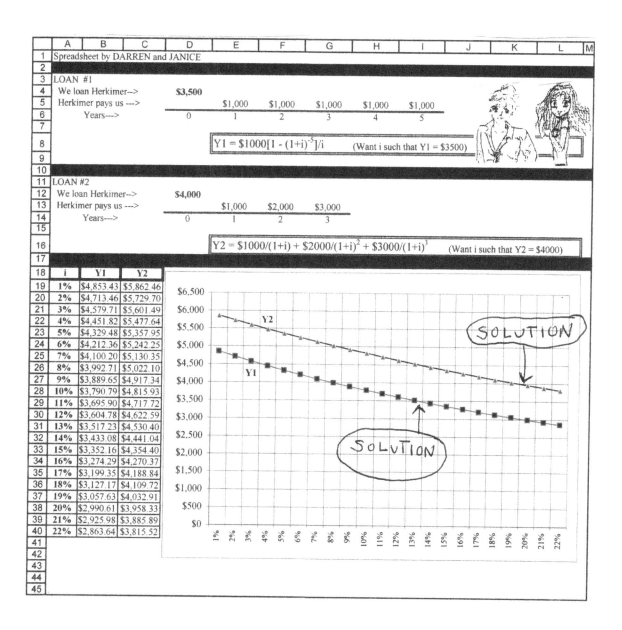

"Neato job," said Herkimer. "If you don't need solutions accurate to two decimal places, a graphic representation is sometimes more readable than tables of numbers."

HERKIMER'S FASCINATING FINANCIAL FACTS:

On the back of a $1 bill on the left side, there is a pyramid. It is a symbol representing material strength and endurance. The pyramid is unfinished. This represents a striving towards growth and a goal of perfection. Above the unfinished pyramid is an eye inside a triangle surrounded by light, representing the eternal eye of God and the theme that the spiritual is above the material. Above the eye the 13-letter Latin motto *Annuit Coeptis* means "He has favored our undertakings." The base of the pyramid contains the Roman numerals MDCCLXXVI (1776). Below the base, the Latin motto *Novus Ordo Seclorum* means "a new order of the ages."

ACTIVITY SET FOR SESSION #13

1. **ALCAL** activity: Use your calculator to check the following values that appear in the spreadsheet table produced by Darren and Janice:

   (a) Present value of loan #1 at 8%:

To be evaluated	Calculator entry	Number represented
$1000[1-(1.08)^{-5}]/.08$	1000*(1-1.08^5)/.08	

   (b) Present value of loan #2 at 7%:

To be evaluated	$1000(1.07)^{-1}+2000(1.07)^{-2}+3000(1.07)^{-3}$
Calculator entry	1000*1.07^1+2000*1.07^2+3000*1.07^3
Number represented	

2. REFLECTION & COMMENT activity:

   Many people invest without adequately learning about the investment process or the different investment products and without considering what they really want to achieve over the long term. These kinds of investors often react to the short-term vicissitudes of the markets, heed the advice of self-proclaimed gurus, buy "hot stocks" at exactly the wrong time, and subsequently end up never getting ahead.

   (John Rogers, CFA President, 2009)

3. A deposit of $20,000 now allows one to withdraw year-end payments of $1400 for 25 years. Find the annual yield on this investment.

4. To pay off a loan of $10,000 an individual makes year-end payments of $1700 for 10 years. What is the annual yield earned by the lender?

5. A deposit of $50,000 now will allow one to withdraw $40,000 in 5 years and $80,000 in 10 years. Calculate the annual yield on this investment.

6. You lend me $1200. To pay off this loan, I will provide you a total of $2000 over a four-year period. Here are three payment options. Find the annual yield for each option. As the lender, which option would you prefer?

OPTION 1: I will pay you $500 at the end of each year for 4 years.

Loan $1200

Payments	.	$500	$500	$500	$500
Year	0	1	2	3	4

OPTION 2: I will pay you $1000 after 2 years and another $1000 after 4 years.

Loan $1200

Payments	.		$1000		$1000
Year	0	1	2	3	4

OPTION 3: I will pay you $2000 after 4 years.

Loan $1200

Payments	.				$2000
Year	0	1	2	3	4

$$
$$

Session #14
THE PACK PRODUCES MORE USEFUL FINANCIAL TABLES

> *To be clever enough to get all the money one desires, one must be stupid enough to want it.*
> -Gilbert K. Chesterton

Once again Herkimer wanted to point out to the Pack that financial tables can sometimes be useful in analyzing investments. At the start of the session, he put two models on the chalkboard, the first being a present value model and the second being an accumulation model. In each case i represents an annual interest rate.

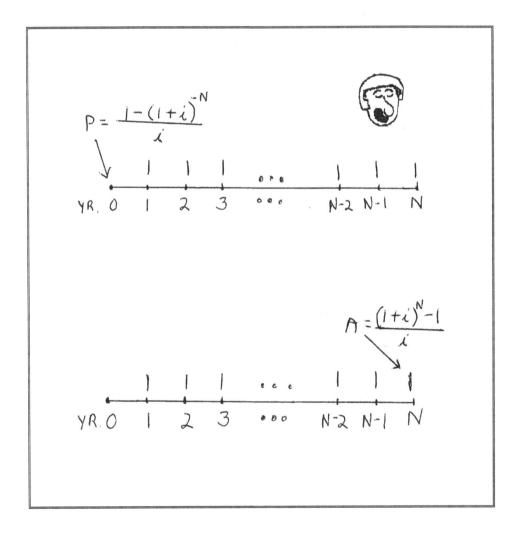

Herkimer asked Pack members to construct a usual financial table for each model along with two examples illustrating use of the tables. The team of Janice and Carolyn came up with these spreadsheet tables:

	A	B	C	D	E	F	G	H	I	J	K	L	M
1	Spreadsheet by		JANICE										
2													
3	**PRESENT VALUE of year-end payments of $1 at various interest rates**												
4	**Table displays values of $P = [1 - (1 + i)^{-n}]/i$ rounded to two decimal places.**												
5													
6													
7			**P**										
8	Pymts			1	1	----------		1	1	1			
9	Yrs		0	1	2	----------		n-2	n-1	n			
10													
11			ANNUAL INTEREST RATE (i)										
12			1%	2%	3%	4%	5%	6%	7%	8%	9%	10%	
13	Y	1	0.99	0.98	0.97	0.96	0.95	0.94	0.93	0.93	0.92	0.91	
14	E	2	1.97	1.94	1.91	1.89	1.86	1.83	1.81	1.78	1.76	1.74	
15	A	3	2.94	2.88	2.83	2.78	2.72	2.67	2.62	2.58	2.53	2.49	
16	R	4	3.90	3.81	3.72	3.63	3.55	3.47	3.39	3.31	3.24	3.17	
17	S	5	4.85	4.71	4.58	4.45	4.33	4.21	4.10	3.99	3.89	3.79	
18		6	5.80	5.60	5.42	5.24	5.08	4.92	4.77	4.62	4.49	4.36	
19		7	6.73	6.47	6.23	6.00	5.79	5.58	5.39	5.21	5.03	4.87	
20		8	7.65	7.33	7.02	6.73	6.46	6.21	5.97	5.75	5.53	5.33	
21		9	8.57	8.16	7.79	7.44	7.11	6.80	6.52	6.25	6.00	5.76	
22		10	9.47	8.98	8.53	8.11	7.72	7.36	7.02	6.71	6.42	6.14	
23		11	10.37	9.79	9.25	8.76	8.31	7.89	7.50	7.14	6.81	6.50	
24		12	11.26	10.58	9.95	9.39	8.86	8.38	7.94	7.54	7.16	6.81	
25		13	12.13	11.35	10.63	9.99	9.39	8.85	8.36	7.90	7.49	7.10	
26		14	13.00	12.11	11.30	10.56	9.90	9.29	8.75	8.24	7.79	7.37	
27		15	13.87	12.85	11.94	11.12	10.38	9.71	9.11	8.56	8.06	7.61	
28		16	14.72	13.58	12.56	11.65	10.84	10.11	9.45	8.85	8.31	7.82	
29		17	15.56	14.29	13.17	12.17	11.27	10.48	9.76	9.12	8.54	8.02	
30		18	16.40	14.99	13.75	12.66	11.69	10.83	10.06	9.37	8.76	8.20	
31		19	17.23	15.68	14.32	13.13	12.09	11.16	10.34	9.60	8.95	8.36	
32		20	18.05	16.35	14.88	13.59	12.46	11.47	10.59	9.82	9.13	8.51	
33													
34													
35		Examples of table use:											
36													
37		(1)	If you invested $11.69 now at an annual interest rate of 5%, you could										
38			withdraw $1 a year at the end of each year for 18 years. In other words,										
39			an investment of $11.69 would allow you to withdraw a total of $18 in										
40			the manner described.										
41													
42		(2)	If you wanted to withdraw year-end payment of $1000 at the end of										
43			each year for the next 20 years a deposit of $1000(9.82) = $9820 now would										
44			allow for this if the annual interest rate is 8%. That is, an investment of										
45			$9820 would allow you to withdraw a total of $20,000 in the manner described.										

	A	B	C	D	E	F	G	H	I	J	K	L	M
1	Spreadsheet by		CAROLYN										
2													
3	**ACCUMULATED VALUE of year-end payments of \$1 at various interest rates**												
4	**Table values values of $A = [(1 + i)^n - 1]/i$ rounded to two decimal places.**												
5													
6										**A**			
7	Pymts			1	1	----------			1	1	1		
8	Yrs		0	1	2	----------		n-2	n-1	n			
9													
10			ANNUAL INTEREST RATE										
11			**1%**	**2%**	**3%**	**4%**	**5%**	**6%**	**7%**	**8%**	**9%**	**10%**	
12	Y	**1**	1.00	1.00	1.00	1.00	1.00	1.00	1.00	1.00	1.00	1.00	
13	E	**2**	2.01	2.02	2.03	2.04	2.05	2.06	2.07	2.08	2.09	2.10	
14	A	**3**	3.03	3.06	3.09	3.12	3.15	3.18	3.21	3.25	3.28	3.31	
15	R	**4**	4.06	4.12	4.18	4.25	4.31	4.37	4.44	4.51	4.57	4.64	
16	S	**5**	5.10	5.20	5.31	5.42	5.53	5.64	5.75	5.87	5.98	6.11	
17		**6**	6.15	6.31	6.47	6.63	6.80	6.98	7.15	7.34	7.52	7.72	
18		**7**	7.21	7.43	7.66	7.90	8.14	8.39	8.65	8.92	9.20	9.49	
19		**8**	8.29	8.58	8.89	9.21	9.55	9.90	10.26	10.64	11.03	11.44	
20		**9**	9.37	9.75	10.16	10.58	11.03	11.49	11.98	12.49	13.02	13.58	
21		**10**	10.46	10.95	11.46	12.01	12.58	13.18	13.82	14.49	15.19	15.94	
22		**11**	11.57	12.17	12.81	13.49	14.21	14.97	15.78	16.65	17.56	18.53	
23		**12**	12.68	13.41	14.19	15.03	15.92	16.87	17.89	18.98	20.14	21.38	
24		**13**	13.81	14.68	15.62	16.63	17.71	18.88	20.14	21.50	22.95	24.52	
25		**14**	14.95	15.97	17.09	18.29	19.60	21.02	22.55	24.21	26.02	27.97	
26		**15**	16.10	17.29	18.60	20.02	21.58	23.28	25.13	27.15	29.36	31.77	
27		**16**	17.26	18.64	20.16	21.82	23.66	25.67	27.89	30.32	33.00	35.95	
28		**17**	18.43	20.01	21.76	23.70	25.84	28.21	30.84	33.75	36.97	40.54	
29		**18**	19.61	21.41	23.41	25.65	28.13	30.91	34.00	37.45	41.30	45.60	
30		**19**	20.81	22.84	25.12	27.67	30.54	33.76	37.38	41.45	46.02	51.16	
31		**20**	22.02	24.30	26.87	29.78	33.07	36.79	41.00	45.76	51.16	57.27	
32													
33													
34	Examples of table use:												
35													
36		(1)	If you invested \$1 at the end of each year for 15 years at an annual										
37			interest rate of 7%, you would have \$25.13 at the end of the 15th year. The										
38			total of \$15 invested as described would yield \$10.13 in interest.										
39													
40		(2)	If you invested \$10,000 at the end of each year for 10 years at an										
41			annual rate of 10%, you would have \$10,000(15.94) = \$159,400 at the end										
42			of the 10th year. The total of \$100,000 invested as described would										
43			produce \$59,400 in interest.										

Herkimer complimented Janice and Carolyn for the neatly-constructed tables. He then had the Pack do some simple exercises to be sure they understood how to use the tables.

ACTIVITY SET FOR SESSON #14

1. **ALCAL** activity:

(a) Use your calculator to check the following value that appears in the spreadsheet table produced by Janice: The present value of year-end payments of $1 for 20 years at 10%:

To be evaluated	Calculator entry	Number represented
$[1-(1.10)^{-20}]/.10$	(1-1.10^~20)/.10	

(b) Use your calculator to check the following value that appears in the spreadsheet table produced by Carolyn: The accumulated value of year-end payments of $1 for 20 years at 10%:

To be evaluated	Calculator entry	Number represented
$[(1.10)^{20}-1]/.10$	(1.10^20-1)/.10	

2. REFLECTION & COMMENT activity:

Impulsive spending completely eliminates the possibility of increasing the value of money over time. Did you know that 85 percent of all Americans who win lotteries spend every penny of their winnings on consumable goods rather than investing in high-yield programs? Based on this statistic, it's plain that the majority of Americans do not understand the profound power of the time/value of money and are destroying their future because of it.

.

(*Money Mastery*, by Williams, Jeppson, and Botkin)

3. Use the PRESENT VALUE table constructed by Janice to respond to the following questions. In each case, determine the amount of interest earned by the deposit described.

   (a) If the annual interest rate is 8%, what deposit would be required now to be able to withdraw payments of $10,000 at the end of each year for 16 years?

   (b) What is the present value of future year end payments of $1,000 for 20 years if the annual interest rate is 7%?

   (c) What deposit earning an annual rate of 6% would be required now to withdraw payments of $100,000 at the end of each year for 10 years?

4. Use the ACCUMULATED VALUE table constructed by Carolyn to respond to the following requests. In each case, find the amount of interest earned by the deposits described.

   (a) If you invested $10,000 at the end of each year for 18 years, what amount would you have at the end of the 18th year if the deposits earned an annual yield of 8%?

   (b) What would be the accumulated value after 20 years of end-of-year deposits of $100,000 if the deposits earned an annual yield of 7%?

   (c) If the annual interest rate is 6%, what would be the accumulation of 15 end-of-year deposits of $1,000 at the end of the 15th year?

5. Consider an annual interest rate i. If P is the present value of n year-end payments of $1 and A is the accumulated value of these payments, show algebraically that $P(1+i)^n = A$.

$$\$$$
$$\$$$

> *Money isn't everything, but it sure keeps you in touch with your children.*
> -J. Paul Getty

Herkimer loved to demonstrate the beauty of algebra. He put this transparency on the overhead and gave the Pack time to look it over:

Deposit		**P**		D		D	D				D	**A**
Y = End of Year	0	1	2	3	4	5	6	7	8	9	10	

D = deposit.

P = present value of deposits (at Y = 0) at annual interest rate i.

A = accumulated value of deposits (at Y = 10) at annual interest rate i.

Above ==> $P(1+i)^{10} = A$.

$P = D(1+i)^{-2} + D(1+i)^{-4} + D(1+i)^{-5} + D(1+i)^{-9}$.

$A = D(1+i)^{8} + D(1+i)^{6} + D(1+i)^{5} + D(1+i)$.

Check out this algebra:

$$\mathbf{P(1+i)^{10}} = \left[D(1+i)^{-2} + D(1+i)^{-4} + D(1+i)^{-5} + D(1+i)^{-9}\right](1+i)^{10}$$

$$= D(1+i)^{-2}\mathbf{(1+i)^{10}} + D(1+i)^{-4}\mathbf{(1+i)^{10}} + D(1+i)^{-5}\mathbf{(1+i)^{10}} + D(1+i)^{-9}\mathbf{(1+i)^{10}}$$

$$= D(1+i)^{8} + D(1+i)^{6} + D(1+i)^{5} + D(1+i)$$

$$= \mathbf{A}.$$

"Wow!" exclaimed Valarie, "the algebra really pulls it all together."

"You are indeed correct," replied Herkimer. "Now I'd like to see you do a similar presentation with the formulas relating to this series of payments assuming an annual interest rate i.

	P						**A**
Deposit		D	D	------	D	D	D.
Y = End of Year	0	1	2	------	n-2	n-1	n

A few minutes passed. The team of Frances and Stephen went to the chalkboard and wrote the following.

$$P = D\left[\frac{1-(1+i)^{-N}}{i}\right]$$

$$A = D\left[\frac{(1+i)^N - 1}{i}\right]$$

$$P(1+i)^N = D\left[\frac{1-(1+i)^{-N}}{i}\right](1+i)^N$$

$$= D\left[\frac{(1+i)^N - (1+i)^0}{i}\right]$$

$$= D\left[\frac{(1+i)^N - 1}{i}\right]$$

$$= A.$$

"Very good," said Herkimer. "Let's look at some of the algebra involved in that chalkboard work." He put this display on the overhead:

82

$$a(b + c) = ab + ac.$$

$$(a/b)c = ac/b.$$

$$a^{-1} = 1/a \text{ if } a \neq 0.$$

$$a^0 = 1 \text{ if } a \neq 0.$$

$$a^x a^{-x} = 1 \text{ if } a \neq 0.$$

Pack members were really starting to appreciate the fact that they had had a solid background in algebra.

HERKIMER'S FASCINATING FINANCIAL FACTS:

In March, 2008, the U.S. Department of the Treasury released new $5 bills into circulation. These bills have some very interesting features including a large purple "5" in the bottom right corner on the back side of the bill and a multitude of small yellow "5" numerals above and to the left of the purple "5." (You have to look closely to see the yellow 5's, but they are there.)

ACTIVITY SET FOR SESSION #15

1. **ALCAL** activity: Consider year end deposits of $1000 for 50 years at annual rate 8%:

(a) Calculate present value of the payments:

To be evaluated	Calculator entry	Number represented
$1000[1-(1.08)^{-50}]/.08$	$1000*(1-1.08^{\wedge}50)/.08$	

(b) Calculate accumulated value of the payments at end of 50th year:

To be evaluated	Calculator entry	Number represented
$1000[(1.08)^{50}-1]/.08$	$1000*(1.08^{\wedge}50-1)/.08$	

(c) Show that the present value calculated in (a) multiplied by $(1.08)^{50}$ is equal to the accumulated value calculated in (b).

2. REFLECTION & COMMENT activity:

The lottery is a tax on the poor, and on people who can't do math. I'm not riding a moral high horse. Research shows that people from lower income brackets, folks who can't afford to be throwing away their money on some ridiculous game, spend four times as much on lottery tickets as anyone else. Rich people don't mess with this garbage, because they know the lottery isn't a wealth-building tool. When was the last time you saw a line of BMW's or Mercedes' pulled up to your local convenience store to buy lottery tickets?

(Dave Ramsay, author of *The Total Money Makeover*)

3. If the annual yield is 7.4%, find the present value (**P**) of the displayed deposits, the accumulated value (**A**) of the deposits at Y = 8, and verify that $P(1.074)^8 = A$.

(a)

	P								**A**
Deposit		$1000	$1000	$1000	$1000	$1000	$1000	$1000	$1000
Y = End of Year	0	1	2	3	4	5	6	7	8

(b)

	P								**A**
Deposit	$1000	$1000	$1000	$1000	$1000	$1000	$1000	$1000	$1000
Y = End of Year	0	1	2	3	4	5	6	7	8

(c)

	P								**A**
Deposit	$1000	$1000	$1000	$1000	$1000	$1000	$1000	$1000	.
Y = End of Year	0	1	2	3	4	5	6	7	8

4. If the annual yield is 6.9%, find the present value (**P**) of the displayed deposits, the accumulated value (**A**) of the deposits at Y = 6, and verify that $P(1.069)^6 = A$.

(a)

	P						**A**
Deposit			$10,000		$20,000		.
Y = End of Year	0	1	2	3	4	5	6

(b)

	P						**A**
Deposit		$1000	$2000			$5000	.
Y = End of Year	0	1	2	3	4	5	6

$$\$$$
$$\$$$

84

<table>
<tr><td colspan="2">HERKIMER'S QUICK QUIZ
SESSIONS 11-15.

Find the correct response for
each question in the 25 cells
at the bottom of the page.</td><td></td></tr>
</table>

Questions 1-4 reference this diagram

Deposit			$5000	$6000			$4000			
Year	0	1	2	3	4	5	6	7	8

1. If the annual interest rate is 5.83%, what is the accumulated value of the displayed deposits at year 8?

2. If the annual interest rate is 5.83%, what is the present value of the displayed deposits at year 0?

3. If the displayed deposits accumulate to $22,000 at year 8, what annual interest rate is earned?

4. If a deposit of $9000 at year 0 allows the withdrawal of the displayed payments, what annual interest rate is earned if nothing remains after the third withdrawal?

5. If you borrow $20,000 and pay it back with year end payments of $5000 for 6 years, what annual yield is earned by the lender?

6. What equal year-end withdrawals can be made for 10 years from a single deposit of $80,000 earning an annual rate of 6% a year?

7. If you wanted to withdraw $15,000 at the end of each year for the next 12 years, what single deposit earning an annual rate of 6.2% would be required at the present time?

8. At an annual rate of 7%, the present value of twenty year-end payments of $2500 for 20 years is _____.

9. At an annual rate of 7%, the present value of twenty beginning-of year payments of $2500 for 20 years is _____.

10. If $1200(1 + i) = $1000(1 + i)^2$, then i = _____.

	A	B	C	D	E
1	$28,339	20.00%	13.22%	$14,286	5.75%
2	$138,651	14.29%	$6,598	$141,345	$26,485
3	8.62%	$54,652	$124,390	23.73%	4.83%
4	$36,845	$12,373	$19,470	9.23%	$25,354
5	11.91%	$10,869	16.83%	$119.432	12.98%

Session #16
THE ANATOMY OF A LOAN

> *Ever notice how it's a penny for your thoughts, yet you put in your two-cents worth?*
> *Someone is making a penny on the deal.*
> -Steven Wright

"We have arrived," said Herkimer excitedly. "We now know enough to discuss and illustrate how loans work. Are you ready to rock and roll?"

The Pack expressed anticipated enthusiasm. Herkimer began by noting that most loans are made on a monthly basis in terms of rates and payments, but that he was going to start out illustrating loans with yearly rates and payments to help the Pack understand basic loan concepts.

"OK," said Herkimer, "here's the story. I loan you $10,000 at a 7% annual rate. You will repay the loan with 4 payments, called *installments*, at the end of each year for the next 4 years. Now I'd like you to study what is displayed on this overhead. It's important that this makes sense to you at this time. The $10,000 is the present value of the future payments you will make to me at 7% annual interest."

Loan = $10,000

| Yr. | 0 | x 1 | x 2 | x 3 | x 4 |

Annual rate i = 7%.

x = installment payment

$10,000 = x[1 - (1.07)^{-4}]/.07 \implies x = $10,000(.07)/[1 - (1.07)^{-4}] = $2952.28.

Interest = 4($2952.28) - $10,000 = $1809.12.

After a brief period of time, Pack members indicated that they were able to understand all of the computations involved on the overhead.

"Now here's a vital point," said Herkimer. "In this simple example, I have made an investment of $10,000 and I want it to yield 7% a year. If you pay me $2952.28 at the end of each year for 4 years, I will accomplish my investment goal. You are paying me a total of 4($2952.28) = $11,809.12 over a 4-year period for the privilege of using my investment of $10,000. Keep in mind that once I lend the $10,000 to you, I no longer have it to spend. I am giving up money at the present time knowing that it will return to me a greater amount in the future. This is a simple example of how money works in a capitalistic society. You

86

need money now, I lend you money I can spare, and you return more than I lent to you. We both benefit."

Herkimer went on to explain that each installment payment of $2952.28 contained interest and principal repayment. *Principal,* he explained, was the portion of the payment that was applied directly to reducing the amount owned. He elaborated on this point: "Since I want to earn 7% a year on my investment of $10,000, the interest I require doing the first year is (0.07)($10,000) = $700. So, your payment of $2952.28 contains $700 of interest and you the remainder of $2952.28 - $700 = $2252.28 is repayment of principal. Hence, after the first payment, you still owe me $10,000 - $2252.28 = $7747.72. This is called the *outstanding principal* after the first installment."

Herkimer asked the Pack to notice that the interest portion of the first installment payment was 700/2952.28 = 0.2371, or approximately 24% of the installment. He went on to explain that the interest required in the second payment would be (0.07)($7747.72) = $542.34. This meant that the principal portion was $2952.28 - $542.34 = $2409.94 and that the outstanding principal after the second payment would be $7747.72 - $2409.94 = $5337.78.

The Pack reflected on this for a few minutes and eventually came to the unanimous conclusion that they understood what was going on. Herkimer then put the entire loan payment schedule on the overhead and asked the group to reflect on what they were observing.

End Year	Payment	Interest	Principal	Principal Outstanding	% interest in payment
0				$10,000.00	
1	$2952.28	(.07)($10,000.00) = $700.00	$2952.28-$700.00= $2252.28	$10,000.00 - $2252.28= $7747.72	23.71%
2	$2952.28	(.07)($7747.72) = $542.34	$2952.28-$532.34= $2409.94	$7747.72 - $2409.94= $5337.78	18.37%
3	$2952.28	(.07)($5337.78) = $373.64	$2952.28-$373.64= $2578.64	$5337.78 - $2578.64 = $2759.14	12.66%
4	$2952.28	(.07)($2759.14) = $193.14	$2952.28-$193.14= $2759.14	$2759.14 - $2759.14 = $0.00	6.54%
SUMS	$11,809.12	$1809.12	$10,000.00		

"My gosh," responded Wayne, "it all fits together. The principal outstanding after the last payment is $0 and the total of the principal column is the loan amount."

"Indeed it does," replied Herkimer. "Now I would like you to realize that there is a very useful formula for calculating the value of the installment payment for a loan if one knows the loan amount, the annual rate, and the duration of the loan. It is easily derived from the present value formula we already know." He put this transparency on the overhead:

87

GENERAL FORMULA FOR INSTALLMENT PAYMENT

Annual rate = i

Loan Amount = **P**

Loan payment		x	x	------	x	x	x .
Y = End of Year	0	1	2	------	n-2	n-1	n

If the annual interest rate is i the single deposit **P** (present value) required to withdraw the displayed year-end payments D is

$$P = x[1-(1+i)^{-n}]/i ==> x = (iP)/[1 - (1+i)^{-n}]$$

The Pack applied this formula to the Herkimer loan situation with **P** = \$10,000, i = 7%, and n = 4. Calculations yielded x = (0.07)(\$10,000)/[1-(1.07)^{-4}] = \$2928.28.

"OK," said Herkimer as he gave a blank transparency to each of five 2-student teams. "I lend you \$50,000 at 9%, and you will repay me with 5 equal year-end payments. Let's see who can produce an installment payment schedule like the ones you have just seen."

It took a while, but the team of Brenda and Roger produced this transparency:

88

"OK, good show," said Herkimer, obviously pleased with the level of understanding demonstrated by Brenda and Roger. "You've noted that the

interest portion of each installment declines as payments are made. And now I'd like a spreadsheet demo for this loan if the annual installments are spread out over 20 years."

The team of Janice and Darren met the challenge.

	A	B	C	D	E	F	G	H
1	Spreadsheet by JANICE and DARREN							
2	A loan of $50,000 paid back over a 20-year period with annual payments.							
3	$50,000	<--- LOAN AMOUNT						
4	9.00%	<---ANNUAL INTEREST RATE						
5	20	<---NUMBER OF YEARS						
6								
7	$5,477.32	<----MONTHLY PAYMENT						
8								
9	Year	Payment	Interest	Principal	Princ.Outstand.		% int. in pymt.	
10	0				$50,000.00			
11	1	$5,477.32	$4,500.00	$977.32	$49,022.68		82.16%	
12	2	$5,477.32	$4,412.04	$1,065.28	$47,957.39		80.55%	
13	3	$5,477.32	$4,316.17	$1,161.16	$46,796.24		78.80%	
14	4	$5,477.32	$4,211.66	$1,265.66	$45,530.57		76.89%	
15	5	$5,477.32	$4,097.75	$1,379.57	$44,151.00		74.81%	
16	6	$5,477.32	$3,973.59	$1,503.73	$42,647.27		72.55%	
17	7	$5,477.32	$3,838.25	$1,639.07	$41,008.20		70.08%	
18	8	$5,477.32	$3,690.74	$1,786.59	$39,221.61		67.38%	
19	9	$5,477.32	$3,529.94	$1,947.38	$37,274.23		64.45%	
20	10	$5,477.32	$3,354.68	$2,122.64	$35,151.59		61.25%	
21	11	$5,477.32	$3,163.64	$2,313.68	$32,837.91		57.76%	
22	12	$5,477.32	$2,955.41	$2,521.91	$30,316.00		53.96%	
23	13	$5,477.32	$2,728.44	$2,748.88	$27,567.11		49.81%	
24	14	$5,477.32	$2,481.04	$2,996.28	$24,570.83		45.30%	
25	15	$5,477.32	$2,211.37	$3,265.95	$21,304.88		40.37%	
26	16	$5,477.32	$1,917.44	$3,559.88	$17,744.99		35.01%	
27	17	$5,477.32	$1,597.05	$3,880.27	$13,864.72		29.16%	
28	18	$5,477.32	$1,247.82	$4,229.50	$9,635.22		22.78%	
29	19	$5,477.32	$867.17	$4,610.15	$5,025.07		15.83%	
30	20	$5,477.32	$452.26	$5,025.07	$0.00		8.26%	
31								
32	TOTALS	$109,546.48	$59,546.48	$50,000.00				
33								
34								

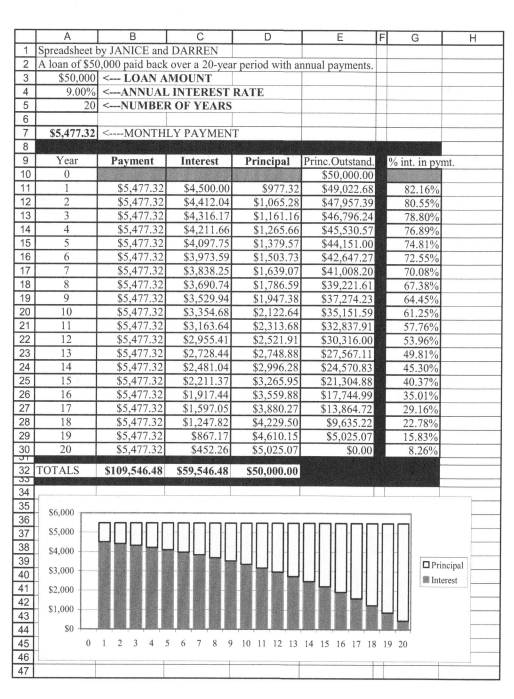

89

"Nifty display," responded Herkimer. "The chart really shows that over half of the total of the loan payments goes to paying off interest."

HERKIMER'S FASCINATING FINANCIAL FACTS:

The legend *In God We Trust* became part of the design of U.S. currency in 1957 and has appeared on all currency since 1963.

ACTIVITY SET FOR SESSION #16

1. **ALCAL** activity:

   (a) Use your calculator to check the annual loan payment calculated on the transparency displayed by Brenda and Roger:

To be evaluated	Calculator entry	Number represented
$.09(50000)/[1-(1.09)^{-5}]$	.09*50000/(1-1.09^5)	

   (b) Use your calculator to check the annual loan payment displayed on the spreadsheet produced by Janice and Darren:

To be evaluated	Calculator entry	Number represented
$.09(50000)/[1-(1.09)^{-20}]$	09*50000/(1-1.09^20)	

2. REFLECTION & COMMENT activity:

   Borrowing money affects your personal credit, no matter the state of the economy. The loan will appear on your credit report and any late payments will negatively affect your credit score. You only want to take out a loan if it's absolutely necessary since you are paying interest on the loan. If you decide to pursue a loan of any kind, it's necessary that you shop around to make sure you're getting the best interest rate.

   (www.savvysugar.com, November 2008)

In activities 3-5 complete the payment schedule for the described loans.

3. A $5000 loan at 12% paid off with 3 equal end-of-year installments.

Yr	Payment	Interest	Principal	Princip. OS	% Interest
0				**$5000.00**	
1					
2					
3					
TOTALS					

4. A $12,000 loan at 8% paid off with 5 equal end-of-year installments.

Yr	Payment	Interest	Principal	Princip. OS	% Interest
0				**$12,000.00**	
1					
2					
3					
4					
5					
TOTALS					

5. A $75,000 loan at 14% paid off with 6 equal end-of-year installments.

Yr	Payment	Interest	Principal	Princip. OS	% Interest
0				**$75,000.00**	
1					
2					
3					
4					
5					
6					
TOTALS					

$$
$$$

91

Session #17
INSTALLMENT PAYMENT FORMULAS ARE SOMETIMES USEFUL

> *Every morning I get up and look through the Forbes list of the richest people in America.*
> *If I'm not there, I go to work.*
> -Robert Orben

Herkimer put the following loan repayment schedule on the overhead. This represented a $30,000 loan at 9% paid off with 6 end-of-year installments. The Pack used the formula developed in the last session to verify the loan payment. They computed $(.09)(\$30.000)/[1-(1.09)^{-6}] = \$6,687.59$.

Y = Year	Payment	Interest	Principal	Princ. Outstanding
0				$30,000.00
1	$6,687.59	$2,700.00	$3,987.59	$26,012.41
2	$6,687.59	$2,341.12	$4,346.48	$21,665.93
3	$6,687.59	$1,949.93	$4,737.66	$16,928.27
4	$6,687.59	$1,523.54	$5,164.05	$11,764.22
5	$6,687.59	$1,058.78	$5,628.81	$6,135.41
6	$6,687.59	$552.19	$6,135.41	$0.00
TOTALS	$40,125.56	$10,125.56	$30,000.00	

There were a number of things Herkimer wanted to indicate to the Pack. The first was that if money figures are rounded to two decimal places, then a table won't necessarily balance out to the nearest penny. For instance, in the above table the year #2 numbers for payment, interest, and principal, expressed to five decimal places, are $6,687.59350, $2,341.11659, and $4,346.47691. The first is the exact sum of the last two (as expected). But the rounded numbers are as shown in the table and the payment is one penny less than the sum of the interest and principal. The difference does not represent an error and is simply due to rounding.

"Now," said Herkimer, "do you realize that the outstanding principal after any payment is simply the present value of the remaining payment? Look, for instance at Y = 3 in the table. After the third payment, the value of $16,928.27 is simply the present value at Y = 3 of the remaining three payments. Check that out, please."

Pack members calculated that $6,687.59[1 - (1.09)^{-3}]/0.09 = \$16,928.26$. They realized the penny difference was due to using a payment figure rounded to the nearest cent.

"Now," said Herkimer, "this will knock your socks off. The principal portions of the payments actually represent a geometric sequence with first term a = $6,687.59/(1.09)^6$ and multiplier $r = 1.09$. Grab your calculators and show this is true."

Sure enough, the Pack discovered that the geometric sequence

$$\$6,687.59/(1.09)^6, \$6,687.59/(1.09)^5, \$6,687.59/(1.09)^4,$$
$$\$6,687.59/(1.09)^3, \$6,687.59/(1.09)^2, \$6,687.59/(1.09)$$

represented the table values $3,987.59, $4,346.48, $4,737.66, $5,164.05, $5,628.81, $6,135.41.

Herkimer went on to emphasize that if one knows the loan installment payment and the principal portion for a payment, then one can easily determine the interest portion of the payment. He also reemphasized that the principal portions increase and interest proportions decrease with each successive payment.

"A payment is a payment. Why is it important to know the interest and principal proportions?" asked Stephen.

"Excellent question," responded Herkimer. "In many situations the interest proportion of a loan repayment represents a tax-deductible item. This is important for people who take out loans to finance the purchase of a home. We'll talk more about this later, but there are definitely situations where knowing the interest portion of a loan repayment can be beneficial when it comes to calculating the tax one owes each year. While most lenders will provide the borrower this information, it's always good to have the math power to check this yourself."

Herkimer decided that it was time to point out that the study to date was primarily with annual payments and annual interest rates. The group had been working with situations where the interest rate period and payment period were the same; basically, a year. He introduced the thought that in many realistic situations, such as home mortgages, the payments are monthly and the loan rate is a yearly rate. "This will be the topic for our next session," he said. "But for now I would like you to produce general formulas for payment, principal, interest, and principal outstanding for the kth installment payment for the loan on this transparency:"

Loan A to be repaid with n year-end installments of P at annual interest rate i.

Loan = A							
Payments	P	P	...	P	...	P	P .
0	1	2	...	k	...	n-1	n

As he frequently did, Herkimer gave the Pack teams a transparency and a transparency pen so that their work could easily be displayed for all to see. The team of Roger and Carolyn were first to make a presentation.

INT. RATE = i

FORMULAS FOR K^{Th} INSTALLMENT:

$$\text{PAYMENT} = P = \frac{iA}{1-(1+i)^{-N}}.$$

$$\text{PRINCIPAL} = \frac{P}{(1+i)^{N+1-K}} = P(1+i)^{-(N+1-K)}.$$

$$\text{INTEREST} = P - P(1+i)^{-(N+1-K)}.$$

PRINCIPAL OUTSTANDING AFTER PAYMENT

$$= \frac{P\left[1-(1+i)^{-(N-K)}\right]}{i}$$

(See Appendix A for algebraic derivation of PRINCIPAL formula.)

"Oh my, that is absolutely gorgeous," said Herkimer. "I couldn't have done it any better. Let's test the beautiful work of Carolyn and Roger with the loan we discussed at the beginning of the session."

The loan was $30,000, the annual rate was 9% and the time period was 6 years. The Pack decided to check the values in the 4th payment. Hence k = 4, n = 6, i = 9% and P = $6,687.59. According to the formulas, the principal in the 4th payment should be

$$\$6{,}687.59(1.09)^{-(6+1-4)} = \$6{,}687.59(1.09)^{-3} = \$5{,}164.05.$$

The interest in the payment should be

$$\$6{,}687.59 - \$5{,}164.05 = \$1{,}523.54.$$

The outstanding principal after the payment should be

$$\$6687.59[1-(1.09)^{-(6-4)}]/.09 = \$6687.59[1-(1.09)^{-2}]/.09 = \$11{,}764.21.$$

"Bingo on all computations," said Herkimer. "Now let me show you a more detailed example of what we have discussed." He put this transparency on the overhead. "Among other things, this should illustrate that the principal portion of each payment increases with each successive payment and the interest portion decreases. This is important for those who take out loans to understand."

Loan = \$800,000, annual rate = 8%.
Loan will be repaid with 30 equal end-of year payments.
Each payment = (\$800,000)(0.08)/[1 - $(1.08)^{-30}$] = \$71,061.95.

Pymt #	Payment	Principal	Interest	Princ. outstanding after payment
2	\$71,061.95	$71,061.95(1.08)^{-(30+1-2)}$ = $71,061.95(1.08)^{-29}$ = \$7,626.90	\$71,061.95 - \$7,626.90 = \$63,435.05	$71,061.95[1-(1.08)^{-(30-2)}]/.08$ = $71,061.95[1-(1.08)^{-28}]/.08$ = \$785,311.19
15	\$71,061.95	$71,061.95(1.08)^{-(30+1-15)}$ = $71,061.95(1.08)^{-16}$ = \$20,742.31	\$71,061.95 - \$20,742.31 = \$50,319.64	$71,061.95[1-(1.08)^{-(30-15)}]/.08$ = $71,061.95[1-(1.08)^{-15}]/.08$ = \$608,253.25
28	\$71,061.95	$71,061.95(1.08)^{-(30+1-28)}$ = $71,061.95(1.08)^{-3}$ = \$56,411.27	\$71,061.95 - \$56,411.27 = \$14,650.68	$71,061.95[1-(1.08)^{-(30-28)}]/.08$ = $71,061.95[1-(1.08)^{-2}]/.08$ = \$126,722.27

The Pack clearly saw that early payments were mostly interest.

HERKIMER'S FASCINATING FINANCIAL FACTS:

Alexander Hamilton (\$10) and Benjamin Franklin (\$100) are the only non-presidents featured on U.S currency. Hamilton, who was born in the West Indies, is the only featured person not born in the United States. He is also the only individual presently featured on currency who faces to the left.

ACTIVITY SET FOR SESSION #17

1. **ALCAL** activity: Relating to the transparency Herkimer placed on the overhead relating to a 30-year loan of \$800,000, use your calculator to:

(a) Check the principal portion of the 28th payment:

To be evaluated	Calculator entry	Number represented
$71061.95(1.08)^{-(30+1-28)}$	71061.95*1.08^-(30+1-28)	

(b) Check the principal outstanding after the 28th payment:

To be evaluated	$71061.95[1-(1.08)^{-(30-28)}]/.08$
Calculator entry	71061.95*(1-1.08^-(30-28))/.08
Number represented	

95

2. REFLECTION & COMMENT activity:

Securing financing to invest in a second home was relatively easy just three to four years ago. Nowadays, however, coming up with the financial means to snag a vacation home is next to impossible - unless you have oodles of cash burning a hole in your pocket. Lenders are hurting, and they're passing their financial pain along to customers.

(Tom Kerr, *Buying a Second Home Becomes More Challenging*, www.mortgageloan.com, June 2009)

3. A loan of \$50,000 will be repaid with equal year-end installments for 10 years. If the annual interest rate is 12%, find the principal and interest portions of (a) the 3rd payment; (b) the 9th payment. In each case, find the principal outstanding after the payment.

4. A loan of \$20,000,000 is to be repaid with year-end installments for 50 years. If the annual interest rate is 8.5%, fill in the appropriate numbers in the table below.

Pymt #	Payment	Principal	Interest	Principal outstanding after payment
3				
12				
37				
48				

$$
$$$

96

> *A bank is a place that will lend you money if you can prove you don't need it.*
> -Bob Hope

Herkimer placed this transparency displaying a bank ad on the overhead:

HERKIMER'S BANK AND TRUST
Two-Year Certificates of Deposits earn

| 6.00% * | 6.167% APR |

* Annual rate compounded monthly

He had previously asked Pack members to look for financial ads where interest rates were listed. A few had done so and did have examples similar to the overhead display where two numerically different interest rates were listed, both relating to the same investment.

Herkimer went on to explain that their work to date had been with annual interest rates. This allowed them to gain an understanding about how compound interest works. But financial schedules often reflect monthly payments and one can't directly apply true annual rates to monthly payments.

"OK," he said, "in the advertisement you are being told that your investment will earn a true annual percentage rate (APR) of 6.167%. So, if you invest $12,000 now, what can you expect to have after 1 year? Then calculate what would you have after 5 years?"

This was a piece of cake for the Pack. They quickly calculated that the investment would grow to $12,000(1.06167) = $12,740.04 after 1 year, and to $12,000(1.06167)^5 = $16,185.61 after 5 years.

"That's fine, and indeed that is accurate," responded Herkimer. "The 6.167% is the *true* annual rate (sometimes called the *effective* rate) of interest. The *true* annual rate is sometimes referenced as the *effective* rate, the *annual yield*, or the *APR* (annual percentage rate). Up to this point, we have been working exclusively with true annual rates. But now let's look at the 6.00%. Note that the asterisk indicates that this is an annual rate *compounded monthly*. The 6.00% annual rate compounded monthly is **not** a true annual rate. Now this doesn't mean it is a false rate." Herkimer chuckled at this point. He then indicated that a true annual rate is compounded just once a year. A rate that is compounded more than once is a *nominal rate*. In this situation, the 6.00% is compounded 12 times a year since it is compounded monthly.

"Of course we have to know what this means," Herkimer continued. "In this case it means that we have a true monthly of 0.06/12 = 0.005 = 0.5%, or 1/2 % a month." Using the chalkboard, he indicated that after 1 month the investment value would be $12,000(1.005) = $12,060, after two months $12,000(1.005)^2 = $12,120.30, after three months $12,000(1.005)^3 = $12,180.90, etc. And, after 12 months, the value would be $12,000(1.005)^{12} = $12,740.13.

"Hey," said Glen, "that's just a few cents off from the value we obtained using the true annual rate of 6.167%."

"Great observation," replied Herkimer. "A true annual rate of 6.167% is *equivalent* to a nominal annual rate of 6% compounded monthly. In general, two rates are *equivalent* if they yield the same yearly interest on a specific investment. Now I'd like to introduce some useful notation relating to nominal rates."

Using the chalkboard, Herkimer wrote $i^{(12)}$ = 6%. He explained that since the 12 was within parenthesis, it was NOT an exponent. In this case it was an indication that the 6% was an annual rate compounded monthly. If one wanted to take a rate i and raise it to the 12th power, one would write i^{12}. The expression $i^{(12)}$ saves a lot of word writing. While $i^{(12)}$ can be read "i upper 12," it simply indicates an annual interest rate compounded monthly. "Now," said Herkimer, "can you accumulate $12,000 for 5 years at $i^{(12)}$ = 6%?"

"I think I get it," responded Valarie. "There are 12(5) = 60 months in 5 years and the true interest rate per month is 1/2%, so the accumulation would be $12,000(1.005)^{60}$, whatever that is." Calculators produced a value of $16,186.20, which was within one dollar of $12,000(1.06167)^5$, previously calculated. Valarie continued, "Yes, I do get it. The rates i= 6.167% and $i^{(12)}$ = 6.00% are equivalent. That is basically what the ad for Herkimer's Bank and Trust says."

"You got it," chirped Herkimer. "And I would like to point out that most ads will give display rates to 2 decimal places. If this had been done in my ad, the rate i would be listed as 6.16%. Rounding up would produce 6.17% and this is more than the actual rate. We are going to spend more time on nominal rates, but for now I'd like to see what you can do with these three problems." He gave Pack teams overhead transparency sheets with the typed problems. They were to use their calculators and produce hand-written solutions. It was requested that they clearly show their written work. Herkimer emphasized to the Pack that the second and third problems dealt only with a single payment, not a series of payments. He also stated that the symbol i will now represent a true annual rate, unless otherwise indicated.

The problems took a bit of time since the concept of a nominal rate was previously not familiar to any of the Pack. After about 10 minutes, the team of Frances and Roger handed Herkimer their transparency sheet. He displayed it for the group:

1. Find the annual yield (APR) that is equivalent to these nominal rates.

   (a) $i^{(12)} = 4.85\%$ (b) $i^{(12)} = 12.32\%$

   (a) $1 + i = \left(1 + \frac{.0485}{12}\right)^{12} \Rightarrow i = \left(1 + \frac{.0485}{12}\right)^{12} - 1 = 4.959\%$

   (b) $1 + i = \left(1 + \frac{.1232}{12}\right)^{12} \Rightarrow i = \left(1 + \frac{.1232}{12}\right)^{12} - 1 = 13.040\%$

2. A single deposit of \$50,000 is made now. Find the value of this investment after 20 years if the annual rate is

   (a) $i = 8.2\%$? (b) $i^{(12)} = 8.2\%$?

   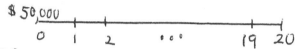

   (a) $\$50,000(1.082)^{20} = \$241,832.81.$

   (b) $\$50,000\left(1 + \frac{.082}{12}\right)^{240} = \$256,324.72.$

3. I want to have \$100,000 available in 10 years. What single deposit must I make now to accumulate to this amount if the annual rate is

   (a) $i = 7.47\%$? (b) $i^{(12)} = 7.47\%$?

   (a) $X(1.0747)^{10} = \$100,000 \Rightarrow X = \frac{\$100,000}{(1.0747)^{10}} = \$48,655.$

   (b) $X\left(1 + \frac{.0747}{12}\right)^{120} = \$100,000 \Rightarrow X = \frac{\$100,000}{\left(1 + \frac{.0747}{12}\right)^{120}} = \$47,488.$

"Oh, that is gorgeous," concluded Herkimer. "What a nifty display. We are definitely ready to move on."

HERKIMER'S FASCINATING FINANCIAL FACTS:

In the 2004 redesigns on paper money, a watermark created during the paper-making process depicts the same historical figure as in the portrait. It is visible from both sides when the bill is held up to a light. On the $20 bill, the watermark of Andrew Jackson is on the right side of the front of the bill.

ACTIVITY SET FOR SESSION #18

1. **ALCAL** activity: Use your calculator to check the annual loan calculations displayed by Frances and Roger on the transparency sheet they produced during the last session:

(a) The annual yield equivalent to $i^{(12)} = 4.85\%$:

To be evaluated	Calculator entry	Number represented
$(1+.0485/12)^{12}-1$	(1+.0485/12)^12-1	

(b) The annual yield equivalent to $i^{(12)} = 12.32\%$:

To be evaluated	Calculator entry	Number represented
$(1+.1232/12)^{12}-1$	(1+.1232/12)^12-1	

2. REFLECTION & COMMENT activity:

Nominal interest rates are not comparable unless their compounding periods are the same; effective interest rates correct for this by "converting" nominal rates into annual compound interest. In many cases, depending on local regulations, interest rates are quoted by lenders and in advertisements are based on nominal, not effective, interest rates, and hence may understate the interest rate compared to the equivalent effective annual rate.

(WikiAnswers.com, 2007)

3. Find the true annual percentage rate (APR) equivalent to each of the displayed $i^{(12)}$ nominal rates:

Nominal rate	$i^{(12)} = 4\%$	$i^{(12)} = 8.33\%$	$i^{(12)} = 14\%$	$i^{(12)} = 22\%$
APR				

4. Note that if $i = 9\%$ and $i^{(12)} = 9\%$, then $i = i^{(12)}$ since the two numbers are equal to 9%. However, the rates are not equivalent. Demonstrate this by calculating the accumulation of $100,000 for 40 years at both rates. (NOTE: **Two numerically equal interest rates are not necessarily equivalent.**)

5. Find the accumulated value of each single-deposit investment.

   (a) $10,000 for 25 years at $i = 9.34\%$.

   (b) $10,000 for 25 years $i^{(12)} = 9.34\%$.

   (c) $600 for 15 years at $i = 17.65\%$.

   (d) $600 for 15 years at $i^{(12)} = 17.65\%$.

6. I want to have $500,000 available in 20 years. What single deposit must I make now to accumulate to this amount if (a) $i = 13.25\%$? (b) $i^{(12)} = 13.25\%$?

7. What is the accumulation of $1000 after one year at each rate?

   (a) A true annual rate of 100%.   (b) $i^{(12)} = 100\%$.

$$\$$$
$$\$$$

> *There is no reason to be the richest man in the cemetery. You can't do any business from there.*
> -Colonel Harland Sanders

Herkimer began this session with this transparency overhead:

HERKIMER'S BANK AND TRUST
One-Year Certificates of Deposits earn

| 5.29% * | 5.43% APR |

Two-Year Certificates of Deposits earn

| 5.73% ** | 5.85% APR |

* Annual rate compounded daily
** Annual rate compounded quarterly

"Build on what you have learned and make some sense out of this ad," was his challenge to the Pack.

The group had no difficulty here. Herkimer verified that in reference to the ad, their notation $i^{(365)} = 5.29\%$ and $i^{(4)} = 5.73\%$ was correct. They also calculated that the annual yield for the daily rate is

$$(1 + .0529/365)^{365} - 1 = 0.05432 \text{ (about 5.43\%)}.$$

For the quarterly rate, the annual yield is

$$(1 + .0573/4)^4 - 1 = 0.05854 \text{ (about 5.85\%)}.$$

These results were consistent with the displayed ad.

"The concept of annual yield is important in the investment world," said Herkimer. "Basically, you want to know what your investment of $1 will be worth in one year. In the above ad, you are being told that $1 will grow to $1.0543 if the annual interest rate is $i^{(4)} = 5.29\%$, and that $1 will grow to $1.0585 if the annual rate is $i^{(365)} = 5.73\%$. Now based on $1, the difference doesn't seem like much. But think about investments involving thousands and millions of dollars. What do you think?"

Herkimer then asked the Pack to assume they were going to invest a large sum of money for a long period of time. If there was a choice between two rates, $i^{(365)} = 7.97\%$ and $i^{(2)} = 8.05\%$, which would they prefer? The alert students realized that the annual yield was important here. Clearly $i^{(365)} < i^{(2)}$, but the Pack used their calculators to determine that the annual yield for the daily rate is 8.29%, while the annual yield for the semiannual rate is 8.21%. In terms of annual yield, the 7.97% is a better rate than 8.05%.

To drive home the concept of nominal rates, Herkimer asked the Pack to accumulate a single deposit of $25,000 for 10 years at each of the following rates:

$i = 12\%$ (True annual rate)	$i^{(2)} = 12\%$ (Annual rate compounded semiannually)	$i^{(4)} = 12\%$ (Annually rate compounded quarterly)	$i^{(12)} = 12\%$ (Annual rate compounded monthly)	$i^{(365)} = 12\%$ (Annual rate compounded daily)

To the nearest dollar, all groups agreed on these answers: $\$25{,}000(1.12)^{10} = \$77{,}646$, $\$25{,}000(1 + .12/2)^{20} = \$80{,}178$, $\$25{,}000(1 + .12/4)^{40} = \$81{,}551$, $\$25{,}000(1 + .12/12)^{120} = \$82{,}510$, and $\$25{,}000(1 + .12/365)^{3650} = \$82{,}987$. It was noted that all annual rates were 12%, but they were not equivalent rates. The more a rate is compounded per year, the better for the investor.

"Now," said Herkimer, "suppose you deposited that $25,000 and it earned an annual rate of 12% compounded every minute. What would you have at the end of the 10 years?"

Pack members calculated that there are $365(24)(60) = 525{,}600$ minutes in a year. Hence, there would be 5,256,000 interest periods over 10 years. They computed
$\$25{,}000(1 + .12/525600)^{5256000} = \$83{,}002.91$

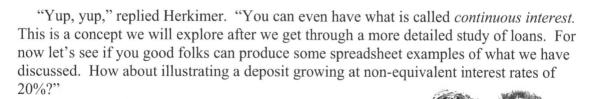

"Holy Herkimer," responded a surprised Darren. "That's not much more than you get if the interest is compounded daily." He then computed what the accumulation would be if the rate of 12% was compounded every second. He found that to be $83,003.58.

"Yup, yup," replied Herkimer. "You can even have what is called *continuous interest*. This is a concept we will explore after we get through a more detailed study of loans. For now let's see if you good folks can produce some spreadsheet examples of what we have discussed. How about illustrating a deposit growing at non-equivalent interest rates of 20%?"

After about ten minutes, Brenda and Glen came up with the following example:

	A	B	C	D	E	F	G	H
1	Spreadsheet by BRENDA and GLEN.							
2								
3	$10,000							
4	**0**	**1**	**2**	**3**	**4**	**5**	<--years	
5								
6	Growth of a single deposit of $10,000 over a 5 year period (to nearest dollar).							
7	Interest rate = 20%.							
8								
9		$i = 20\%$	$i^{(2)} = 20\%$	$i^{(4)} = 20\%$	$i^{(12)} = 20\%$	$i^{(365)} = 20\%$		
10	0	$10,000	$10,000	$10,000	$10,000	$10,000		
11	1	$12,000	$12,100	$12,155	$12,194	$12,213		
12	2	$14,400	$14,641	$14,775	$14,869	$14,917		
13	3	$17,280	$17,716	$17,959	$18,131	$18,218		
14	4	$20,736	$21,436	$21,829	$22,109	$22,251		
15	5	$24,883	$25,937	$26,533	$26,960	$27,175		
16								
17	**APR**	**20.000%**	**21.000%**	**21.551%**	**21.939%**	**22.134%**		
18								
19								
20	**Bars represent, respectively, the accumulation at true rate of 20%, 20% compounded semiannually,**							
21	**20% compounded quarterly, 20% compounded monthly, and 20% compounded daily.**							
22								

"Good show," said Herkimer. "Financial literacy is on the horizon for us. We are just about there."

ACTIVITY SET FOR SESSION #19

1. **ALCAL** activity: Use your calculator to check the following computations on the spreadsheet produced by Brenda and Glen:

   (a) The accumulation of a single deposit of $10,000 for 5 years at $i^{(365)} = 20\%$:

To be evaluated	Calculator entry	Number represented
$10000(1+.20/365)^{(365)(5)}$	10000*(1+.20/365)^(365*5)	

   (b) The annual yield equivalent to $i^{(365)} = 20\%$:

To be evaluated	Calculator entry	Number represented
$(1+.20/365)^{365}-1$	(1+.20/365)^365-1	

2. REFLECTION & COMMENT activity:

   Most savings accounts calculate interest monthly, using a method called annual percentage yield or APY. APY is the amount of interest you earn over a year, but it's a little different than simple interest in that it takes compounding into account. For instance, if your bank offers a 4% per year interest rate on your savings account but compounds your interest monthly, you will actually earn slightly more than 4% on your money over the course of the year.

   (Lindsay Woodland, *About Different Types of Interest Rates*, ehow.com, 2009.)

3. Calculate the annual yield (APR) for each nominal rate.

Rate	$i^{(2)} = 18\%$	$i^{(365)} = 9\%$	$i^{(4)} = 13\%$	$i^{(12)} = 8\%$	$i^{(1000)} = 15\%$
APR					

4. In terms of APR, which of these rates is best for an investor? Which is the worst?

   (a) $i^{(365)} = 10\%$ (b) $i^{(12)} = 10.1\%$ (c) $i^{(4)} = 10.12\%$

5. How much will a single deposit of $100,000 accumulate to after 40 years at each annual rate?

   (a) $i^{(2)} = 9\%$ (b) $i^{(4)} = 9\%$ (c) $i^{(12)} = 9\%$ (d) $i^{(365)} = 9\%$ (e) $i^{(1000)} = 9\%$

6. If you want to have $500,000 available in 40 years, what single deposit would be required to accumulate to this amount if the annual rate is

   (a) $i^{(2)} = 8\%$?   (b) $i^{(4)} = 8\%$?   (c) $i^{(12)} = 8\%$?   (d) $i^{(365)} = 8\%$?   (e) $i^{(1000)} = 8\%$?

7. Find the accumulation of $1000 at the end of one year at each annual rate.

   (a) $i = 100\%$ (b) $i^{(12)} = 100\%$ (c) $i^{(365)} = 100\%$ (d) $i^{(1000)} = 100\%$

$$\$$$
$$\$$$

> *Minutes are worth more than money. Spend them wisely.*
> -Thomas P. Murphy

While explaining that he was about to show one of the most important formulas in financial mathematics, Herkimer put this transparency on the overhead:

If i and $i^{(n)}$ are *equivalent interest rates*, then

$$1 + i = (1 + i^{(n)}/n)^n$$

EXPLANATION:

If \$1 is invested at a *true annual rate* i it will accumulate to $\$1(1 + i)$ after one year.

If n is a positive integer greater than 1 and \$1 is invested at *nominal rate* $i^{(n)}$ it will accumulate to $\$1[(1+i^{(n)}/n]^n$ after one year.

If i and $i^{(n)}$ are *equivalent rates*, then

$$\$1(1 + i) = \$1[(1+i^{(n)})/n]^n$$

$$\Longrightarrow 1 + i = (1 + i^{(n)}/n)^n$$

Note: In the formula above, the number n serves as a divisor, an exponent, and as an indicator to denote the number of times per year that an interest rate is compounded.

He quickly emphasized to the Pack that they had used the formula before to find annual yields (APRs) equivalent to nominal rates. For instance, if $i^{(12)} = 14\%$, then the true annual rate i is found from solving $1 + i = (1 + .14/12)^{12} \Longrightarrow i = (1 + .14/12)^{12} - 1 = 14.934\%$.

Herkimer gave Pack teams a transparency sheet and pen. He asked them to use their algebraic skills to solve the general interest equation for i, then for $i^{(n)}$, and to provide an example using the two solutions. Darren and Brenda were first to present a sheet with Herkimer's request.

$$\boxed{i \text{ and } i^{(N)} \text{ equivalent} \Rightarrow 1 + i = \left(1 + \frac{i^{(N)}}{N}\right)^{N}.}$$

Above $\Rightarrow i = \left(1 + \frac{i^{(N)}}{N}\right)^{N} - 1.$

Example: If $i^{(12)} = 24\%$, Then APR

$= i = \left(1 + \cdot \frac{24}{12}\right)^{12} - 1 = 0.26824 \approx 26.82\%.$

Conclusion: $i^{(12)} = 24\%$ and $i = 26.82\%$ are equivalent rates.

$1 + i = \left(1 + \frac{i^{(N)}}{N}\right)^{N} \Rightarrow 1 + \frac{i^{(N)}}{N} = \left(1 + i\right)^{1/N}$

$$\Rightarrow \frac{i^{(N)}}{N} = \left(1 + i\right)^{1/N} - 1$$

$$\Rightarrow i^{(N)} = N\left[\left(1 + i\right)^{1/N} - 1\right].$$

Example: If $i = 9\%$ and $N = 12$,

$$i^{(12)} = 12\left[\left(1.09\right)^{1/12} - 1\right]$$
$$= 0.0864879$$
$$\approx 8.65\%.$$

Conclusion: $i = 9\%$ and $i^{(12)} = 8.65\%$ are equivalent rates.

"Nice examples," said Herkimer. He then put this transparency on the overhead:

Deposits	$900					$4000								$8000					
Months	0	1	2	3	4	5	6	7	8	9	10	11	12	13	14	15	16	17	18

Find the accumulation of the three deposits after 18 months if they earn

(a) a true annual rate of 10%;   (b) an annual rate $i^{(4)} = 10\%$.

"Oh, I think I see the problem here," said Carolyn. "These are monthly payments and the rates we are given are not monthly rates. We need to find the rate $i^{(12)}$ that is equivalent to those stated in (a) and (b)."

"Neat observation," responded Herkimer. "Go to it."

Carolyn and Glen worked out this presentation on the chalkboard:

DEPOSITS $900 $4000 $8000

MONTHS 0 1 2 3 4 5 6 7 8 9 10 11 12 13 14 15 16 17 18

(a)

$$\left(1 + \frac{i^{(12)}}{12}\right)^{12} = 1.10 \Rightarrow i^{(12)} = 12\left[(1.10)^{1/12} - 1\right] = 9.5689685\%$$

$$\Rightarrow \frac{i^{(12)}}{12} = 0.7974\% \quad (\text{TRUE MONTHLY RATE}).$$

$$\text{ACCUMULATION} = \$900(1.007974)^{18} + \$4000(1.007974)^{13}$$
$$+ \$8000(1.007974)^{5} = \$13,797.48.$$

(b)

$$\left(1 + \frac{i^{(12)}}{12}\right)^{12} = \left(1 + \frac{.10}{4}\right)^{4} \Rightarrow 1 + \frac{i^{(12)}}{12} = \left[\left(1 + \frac{.10}{4}\right)^{4}\right]^{1/12}$$

$$\Rightarrow i^{(12)} = 12\left[\left(1 + \frac{.10}{4}\right)^{1/3} - 1\right] = 9.9178805\%$$

$$\Rightarrow \frac{i^{(12)}}{12} = 0.82648\% \quad (\text{TRUE MONTHLY RATE}).$$

$$\text{ACCUMULATION} = \$900(1.0082648)^{18} + \$4000(1.0082648)^{13}$$
$$+ \$8000(1.0082648)^{5} = \$13,831.57.$$

109

"OK, great work," responded Herkimer. "It's really key to understand that if you have a series of periodic payments and want to accumulate or discount them, then you need an interest rate that corresponds to the payment period. I like to say that the payment period and interest period must *jive* if you want to do accurate computations. One sometimes needs to use the concept of equivalent interest rates as you have demonstrated in the work during this session."

HERKIMER'S FASCINATING FINANCIAL FACTS:

In the years 2001-2004 various members of Congress introduced legislation that would replace the $10 bill portrait of Alexander Hamilton with a portrait of Ronald Reagan. None of the legislation received enough support for passage.

ACTIVITY SET FOR SESSION #20

1. **ALCAL** activity: Use your calculator to check the following computations on the sheet produced by Darren and Brenda:

(a) The APR equivalent to the nominal rate $i^{(12)} = 24\%$:

To be evaluated	Calculator entry	Number represented
$(1+.24/12)^{12}-1$	(1+.24/12)^12-1	

(b) The nominal rate i(12) equivalent to an annual yield of 9%:

To be evaluated	Calculator entry	Number represented
$12[(1.09)^{1/12}-1]$	12*(1.09^(1/12)-1)	

2. REFLECTION & COMMENT activity:

Why do people pay interest? Lenders demand that borrowers pay interest for several important reasons. First, when people lend money, they can no longer use this money to fund their own purchases. The payment of interest makes up for this inconvenience. Second, a borrower may default on the loan. In this case, the borrower fails to pay back the loan and the lender loses the money, less whatever can be recovered from the borrower. Interest helps make the risk of default worth taking. In general, the more risk there is of default on the loan, the higher the interest rate demanded by the lender. Finally, and most importantly, lenders demand interest since while the borrower has the money, inflation tends to reduce the real value, or purchasing power, of the loan. In this case, interest allows the balance due to grow as inflation erodes the real value of the balance due.

(Reasons for Paying Interest, www.sparknotes.com, 2009)

3. Given a true rate $i = 15\%$, find the following equivalent nominal rates:

    (a) $i^{(4)}$ (b) $i^{(12)}$ (c) $i^{(365)}$

4. Deposits of $2000 are made at the beginning of each month for 3 years. Find the accumulated value of these payments at the end of the 3 years if the interest rate is

    (a) $i^{(12)} = 24\%$ (b) $i = 24\%$

5. During a five year period, deposits of $10,000 are made at the end of each of the months 4, 26, 32, 40, and 53. If the APR is 9.6%, find the accumulation of these deposits at the end of the 5th year.

6. A bank advertises that it will pay a rate of 8.2% compounded quarterly on a 9 month certificate of deposit. What is APR on this investment?

7. If an investment has an advertised APR of 7.85%, what are the equivalent annual rates $i^{(4)}$, $i^{(12)}$, and $i^{(365)}$?

$$\$$$
$$\$$$

HERKIMER'S QUICK QUIZ
SESSIONS 16-20.

Find the correct response for
each question in the 25 cells
at the bottom of the page.

1. A ten year $250,000 loan at a true annual rate of 9% would be paid off with ten annual year-end installments of _____.

2. The interest portion in the first installment for the loan described in activity #1 is _____.

3. The principal outstanding after the first installment for the loan described in activity #1 is _____.

4. The principal portion in the second installment for the loan described in activity #1 is _____.

5. The accumulation of a single deposit of $40,000 for 30 years earning an annual rate of $i^{(12)} = 7.5\%$ is _____.

6. What is the APR equivalent to the annual rate $i^{(365)} = 12.5\%$?

7. If the annual interest rate is $i^{(4)} = 6.9\%$, what single deposit made now will accumulate to $30,000 in 12 years?

8. What annual rate $i^{(365)}$ is equivalent to a true annual rate of 22%?

9. A loan of $300,000 will be paid off with year-end installments over 30 years. If the true annual rate is 6.8%, what is the amount of each installment payment?

10. What is the total amount of interest paid for the loan described in activity #9?

	A	B	C	D	E
1	21.08%	$38,955	$56,376	29.65%	13.31%
2	$398,675	14.77%	$15,988	$17,936	$504,843
3	$265,455	24.94%	$233,545	$410,760	$376,861
4	8.59%	$23,692	$13,201	$22,500	$11,743
5	$42,754	$51,244	19.89%	$417,232	16.85%

Session #21
LOANS WITH MONTHLY PAYMENTS

> *Ever wonder about those people who spend $2 apiece on those little bottles of Evian water? Try spelling Evian backward.*
> -George Carlin

When the Pack entered the room for a financial session they found this display on the overhead projector screen:

	Loan (3 years)	Loan (5 years)	Difference
Loan Balance	$16,000	$16,000	
Loan Term (Months)	36	60	
Interest Rate	9%	12%	
Monthly Payment	$508.80	$355.91	$152.89
Total Payments	$18,316.65	$21,354.67	
Total Interest	$2,316.65	$5,354.67	($3,038.02)

After allowing the group time to look over the table, Herkimer said that the table referenced a car loan of $16,000. He stressed the following points:

** The rates are nominal annual rates compounded monthly. That is, $i^{(12)} = 9\%$ on the 3-year loan and $i^{(12)} = 12\%$ on the 5-year loan. Most loan interest rates are nominal annual rates compounded monthly although this is not directly stated. It is implicitly assumed in the real world when annual interest rates are provided.

** In general, the longer the loan period the greater the interest rate. The monthly payments are lower on longer term loans, but one pays more in total interest on the longer term loans. Notice that the monthly payments on the 5-year loan are $152.89 less than those on the 3-year loan, but one pays $3.038.02 more in interest. The parentheses around the number (common on financial statements) indicates that the number is negative.

** The numbers are rounded to the nearest penny. Hence the payment totals may be "pennies off" from the sum of the listed monthly payments. For instance, 36($508.80) = $18,316.80 for the 3-year loan, which differs by 15 cents from the table total.

The Pack had previously worked with loans involving annual installment payments. Herkimer was well aware that most loans involve monthly payments and that most stated loan rates were annual rates compounded monthly. He wanted to be sure Pack members could make the transition from annual payments to monthly payments. He used this chalkboard to make this presentation related to the 3-year loan on the previous display. Prior to the chalkboard work he stated, "We're going to do some spreadsheet work with loans involving monthly payments. It goes without saying that we need to understand how to do the basic computations involved."

113

Herkimer's chalkboard presentation:

$$\text{Loan} = \$16,000.$$

$$\text{Loan Term: } 36 \text{ months.}$$

$$\text{Interest rate: } i^{(12)} = 9\%.$$

$$\text{Monthly payment} = \frac{\$16,000(.09/12)}{1 - (1 + .09/12)^{-36}}$$

$$= \$508.80.$$

$$\text{Total payments} = 36(\$508.80)$$

$$= \$18,316.80.$$

$$\text{Total interest} = \$18,316.80 - \$16,000.00$$

$$= \$2,316.80.$$

Herkimer challenged Pack teams to produce spreadsheets displaying each payment along with the interest and principal portions and the principal outstanding after each payment. He reminded the students that the payments were monthly, and that the true monthly rate could easily be calculated from the nominal annual rates $i^{(12)}$.

The team of Frances and Valarie produced this spreadsheet for the 3-year loan:

	A	B	C	D	E	F	G
1	Spreadsheet by FRANCES and VALARIE						
2							
3	$16,000	<---Amount of Loan					
4	36	<---Loan Term (Months)					
5	0.09	<---Annual Interest Rate (compounded monthly)					
6	0.0075	<---True Monthly Rate					
7							
8	$508.80	<--Monthly payment					
9							% interest
10	Month #	Payment	Interest	Principal	Princ. Outstanding		In Payment
11	0				$16,000.00		
12	1	$508.80	$120.00	$388.80	$15,611.20		23.59%
13	2	$508.80	$117.08	$391.71	$15,219.49		23.01%
14	3	$508.80	$114.15	$394.65	$14,824.84		22.43%
15	4	$508.80	$111.19	$397.61	$14,427.23		21.85%
16	5	$508.80	$108.20	$400.59	$14,026.64		21.27%
17	6	$508.80	$105.20	$403.60	$13,623.05		20.68%
18	7	$508.80	$102.17	$406.62	$13,216.42		20.08%
19	8	$508.80	$99.12	$409.67	$12,806.75		19.48%
20	9	$508.80	$96.05	$412.75	$12,394.01		18.88%
21	10	$508.80	$92.96	$415.84	$11,978.17		18.27%
22	11	$508.80	$89.84	$418.96	$11,559.21		17.66%
23	12	$508.80	$86.69	$422.10	$11,137.10		17.04%
24	13	$508.80	$83.53	$425.27	$10,711.84		16.42%
25	14	$508.80	$80.34	$428.46	$10,283.38		15.79%
26	15	$508.80	$77.13	$431.67	$9,851.71		15.16%
27	16	$508.80	$73.89	$434.91	$9,416.80		14.52%
28	17	$508.80	$70.63	$438.17	$8,978.63		13.88%
29	18	$508.80	$67.34	$441.46	$8,537.18		13.24%
30	19	$508.80	$64.03	$444.77	$8,092.41		12.58%
31	20	$508.80	$60.69	$448.10	$7,644.31		11.93%
32	21	$508.80	$57.33	$451.46	$7,192.84		11.27%
33	22	$508.80	$53.95	$454.85	$6,737.99		10.60%
34	23	$508.80	$50.53	$458.26	$6,279.73		9.93%
35	24	$508.80	$47.10	$461.70	$5,818.03		9.26%
36	25	$508.80	$43.64	$465.16	$5,352.87		8.58%
37	26	$508.80	$40.15	$468.65	$4,884.23		7.89%
38	27	$508.80	$36.63	$472.16	$4,412.06		7.20%
39	28	$508.80	$33.09	$475.71	$3,936.36		6.50%
40	29	$508.80	$29.52	$479.27	$3,457.08		5.80%
41	30	$508.80	$25.93	$482.87	$2,974.22		5.10%
42	31	$508.80	$22.31	$486.49	$2,487.73		4.38%
43	32	$508.80	$18.66	$490.14	$1,997.59		3.67%
44	33	$508.80	$14.98	$493.81	$1,503.77		2.94%
45	34	$508.80	$11.28	$497.52	$1,006.26		2.22%
46	35	$508.80	$7.55	$501.25	$505.01		1.48%
47	36	$508.80	$3.79	$505.01	$0.00		0.74%
48							
49	TOTALS	$18,316.65	$2,316.65	$16,000.00			

Wayne and Darren constructed the following spreadsheet for the 5-year loan:

115

	A	B	C	D	E	F	G
1	Spreadsheet by WAYNE and DARREN						
2							
3	**$16,000**	<---LOAN AMOUNT					
4	5	<--NUMBER OF YEARS					
5	60	<---NUMBER OF MONTHS					
6	12.00%	<---ANNUAL NOMINAL RATE COMPOUNTED MONTHLY					
7	1.00%	<---TRUE MONTHLY RATE					
8							
9	**$355.91**	<--Monthly payment					
10							% INT
11	MONTH #	PYMT	INT	PRINC	PRINC O/S		IN PYMT
12	0				$16,000.00		
13	1	$355.91	$160.00	$195.91	$15,804.09		44.96%
14	2	$355.91	$158.04	$197.87	$15,606.22		44.40%
15	3	$355.91	$156.06	$199.85	$15,406.37		43.85%
16	4	$355.91	$154.06	$201.85	$15,204.52		43.29%
17	5	$355.91	$152.05	$203.87	$15,000.66		42.72%
18	6	$355.91	$150.01	$205.90	$14,794.75		42.15%
19	7	$355.91	$147.95	$207.96	$14,586.79		41.57%
20	8	$355.91	$145.87	$210.04	$14,376.74		40.98%
21	9	$355.91	$143.77	$212.14	$14,164.60		40.39%
22	10	$355.91	$141.65	$214.27	$13,950.34		39.80%
23	11	$355.91	$139.50	$216.41	$13,733.93		39.20%
24	12	$355.91	$137.34	$218.57	$13,515.36		38.59%
25	13	$355.91	$135.15	$220.76	$13,294.60		37.97%
26	14	$355.91	$132.95	$222.97	$13,071.63		37.35%
27	15	$355.91	$130.72	$225.19	$12,846.44		36.73%
28	16	$355.91	$128.46	$227.45	$12,618.99		36.09%
29	17	$355.91	$126.19	$229.72	$12,389.27		35.46%
30	18	$355.91	$123.89	$232.02	$12,157.25		34.81%
31	19	$355.91	$121.57	$234.34	$11,922.91		34.16%
32	20	$355.91	$119.23	$236.68	$11,686.23		33.50%
33	21	$355.91	$116.86	$239.05	$11,447.18		32.83%
34	22	$355.91	$114.47	$241.44	$11,205.74		32.16%
35	23	$355.91	$112.06	$243.85	$10,961.89		31.48%
36	24	$355.91	$109.62	$246.29	$10,715.60		30.80%
37	25	$355.91	$107.16	$248.76	$10,466.84		30.11%
38	26	$355.91	$104.67	$251.24	$10,215.60		29.41%
39	27	$355.91	$102.16	$253.76	$9,961.84		28.70%
40	28	$355.91	$99.62	$256.29	$9,705.55		27.99%
41	29	$355.91	$97.06	$258.86	$9,446.70		27.27%
42	30	$355.91	$94.47	$261.44	$9,185.25		26.54%
43	31	$355.91	$91.85	$264.06	$8,921.19		25.81%
44	32	$355.91	$89.21	$266.70	$8,654.49		25.07%
45	33	$355.91	$86.54	$269.37	$8,385.13		24.32%
46	34	$355.91	$83.85	$272.06	$8,113.07		23.56%
47	35	$355.91	$81.13	$274.78	$7,838.29		22.80%
48	36	$355.91	$78.38	$277.53	$7,560.76		22.02%
49	37	$355.91	$75.61	$280.30	$7,280.46		21.24%
50	38	$355.91	$72.80	$283.11	$6,997.35		20.46%
51	39	$355.91	$69.97	$285.94	$6,711.41		19.66%
52	40	$355.91	$67.11	$288.80	$6,422.61		18.86%
53	41	$355.91	$64.23	$291.69	$6,130.93		18.05%
54	42	$355.91	$61.31	$294.60	$5,836.33		17.23%
55	43	$355.91	$58.36	$297.55	$5,538.78		16.40%
56	44	$355.91	$55.39	$300.52	$5,238.26		15.56%
57	45	$355.91	$52.38	$303.53	$4,934.73		14.72%
58	46	$355.91	$49.35	$306.56	$4,628.16		13.87%
59	47	$355.91	$46.28	$309.63	$4,318.53		13.00%
60	48	$355.91	$43.19	$312.73	$4,005.81		12.13%
61	49	$355.91	$40.06	$315.85	$3,689.95		11.26%
62	50	$355.91	$36.90	$319.01	$3,370.94		10.37%
63	51	$355.91	$33.71	$322.20	$3,048.74		9.47%
64	52	$355.91	$30.49	$325.42	$2,723.32		8.57%
65	53	$355.91	$27.23	$328.68	$2,394.64		7.65%
66	54	$355.91	$23.95	$331.96	$2,062.67		6.73%
67	55	$355.91	$20.63	$335.28	$1,727.39		5.80%
68	56	$355.91	$17.27	$338.64	$1,388.75		4.85%
69	57	$355.91	$13.89	$342.02	$1,046.73		3.90%
70	58	$355.91	$10.47	$345.44	$701.29		2.94%
71	59	$355.91	$7.01	$348.90	$352.39		1.97%
72	60	$355.91	$3.52	$352.39	$0.00		0.99%
73							
74	TOTALS	$21,354.67	$5,354.67	$16,000.00			

116

The students were delighted that their spreadsheet totals "jived" with the figures displayed in the advertisement they saw at the beginning of the session.

Herkimer reminded the Pack that they wouldn't have to print out an entire loan payment schedule to find the interest and principal portions of a payment. "Remember the installment payment formulas we developed in a previous session. They are perfectly good here as long as we remember we are now working with monthly payments and true monthly rates. Let's see who can produce all the details relating to the 50th payment in the 5-year loan illustrated in the spreadsheet produced by Wayne and Darren." [*The formulas appear in Session #17.*]

After a flurry of calculator activity and pencil work, Carolyn and Brenda put this presentation on the chalkboard:

$$Loan = \$16,000.$$
$$Monthly\ Pymt. = \$355.91.$$
$$True\ monthly\ rate = \frac{12\%}{12} = 1\%.$$

For 50th payment

$$Principal = \$355.91(1.01)^{-(60+1-50)}$$
$$= \$355.91(1.01)^{-11}$$
$$= \$319.01.$$

$$Interest = \$355.91 - \$319.01$$
$$= \$36.90.$$

Principal outstanding
$$= \frac{\$355.91[1-(1.01)^{-(60-50)}]}{.01}$$
$$= \$3,370.93.$$

"Excellent, excellent, excellent!" yelped an excited Herkimer. "You young folks are fantastic. Now we are going to move ahead into home mortgages. It's here where the interest portions of payments are really important since interest payments on home loans are tax deductible items. That's not so for interest payments on car loans."

Pack members were excited with Herkimer's enthusiasm. They were ready to move on.

HERKIMER'S FASCINATING FINANCIAL FACTS:

The highest paper money denomination printed in the last 45 years is $100. There is a $100,000 bill, but it was never made available to the public. It was limited to transactions between the Treasury Department and the Federal Reserve. The $100,000 bill is still legal tender if you somehow happen to end up with one.

ACTIVITY SET FOR SESSION #21

1. **ALCAL** activity:

(a) Use your calculator to check the following computation on the
on the spreadsheet produced by Frances and Valarie: The monthly payment on
a 3-year $16,000 loan at $i^{(12)} = 9\%$:

To be evaluated	$16000(.09/12)/[1-(1+.09/12)^{-36}]$
Calculator entry	16000*(.09/12)/(1-(1+.09/12)^36)
Number represented	

(b) Use your calculator to check the following computation on the
on the spreadsheet produced by Wayne and Darren: The monthly payment on
a 5-year $16,000 loan at $i^{(12)} = 12\%$:

To be evaluated	$16000(.12/12)/[1-(1+.12/12)^{-60}]$
Calculator entry	16000*(.12/12)/(1-(1+.12/12)^60)
Number represented	

118

2. REFLECTION & COMMENT activity:

The simplest reason to refinance is to switch to a loan with a lower interest rate. Say you borrowed money to purchase your house at an interest rate of seven percent for a 30-year fixed-rate loan, and today's going rate is six-and-a-half percent. If you refinance, you'll have a lower interest rate over the life of your loan. However, that doesn't necessarily save money in the deal, either over the long term or in your monthly payment. There's more to it than that. Refinancing costs money, and you may be charged for appraisal, origination, and insurance fees, plus title search and legal costs - all of which can add up to thousands of dollars. You may want to consider this rule of thumb: refinance if you can lower your rate by at least one percentage point.

(Joanne Lee, *Refinancing: Make a Good Deal Better*, mortgage.com, 2009)

3. Complete the following table:

Loan Amount	$10,000	$20,000	$25,000	$30,000	$40,000
Loan Term (years)	2	3	4	5	5
Loan Term (months)					
Loan rate $i^{(12)}$	8%	10%	13%	15%	18%
Monthly payment					
Total Payments					
Total Interest Paid					

4. A 9.4% five-year loan of $42,000 will be repaid with monthly payments. Find the monthly payment and the interest and principal portions for the indicated payments along with the principal outstanding after the payment:

   (a) 6th payment   (b) 34th payment   (c) 55th payment

5. Construct the full payment schedule for a 15% one-year loan of $5,000:

Month #	Payment	Interest	Principal	Princ. Outstanding
0				
1				
2				
3				
4				
5				
6				
7				
8				
9				
10				
11				
12				
TOTALS				

6. Research this question: What does it mean to *cosign* for a loan?

$$
$$

Session #22
MORTGAGES - THE BIG LOANS

> *I'd love to go to Washington, if only to be nearer my money.*
> -Bob Hope

"We're moving into the big time," said Herkimer as he prepared to put a transparency on the overhead. "Let me remind you that a loan amount is simply the present value of the future payment that will be received by the lender." The transparency contained the following:

Loan amount = A.
Loan interest rate = $i^{(12)}$.
$I = i^{(12)}/12$ (this is the true monthly rate).
n = number of months (duration of loan).
Monthly payment = P.

```
    A
    .      P      P      P    ---    P      P      P
  ――――――――――――――――――――――――――――――――――――――――――――――――――――
    0      1      2      3    ---   n-2    n-1     n
```

The required monthly payment P is

$$P = IA/[1 - (1 + I)^{-n}]$$

Herkimer explained that housing values vary considerably throughout the country. A house valued at $200,000 in one area might have a value of $800,000 in another section of the United States. He provided an example of a mortgage: A house is sold for $400,000. The buyer puts down 20% ($80,000) of the amount and takes out a 30-year loan (mortgage) at 6% to pay for the remaining $320,000. Remember that loan interest rates are annual rates compounded monthly. In this example $I = 0.06/12 = 0.005$. The borrower will have $12(30) = 360$ monthly payments of

$$(.06/12)(\$320,000)/[1 - (1+.06/12)^{-360}] = \$1,918.56$$

Over the 30 year period, the total of the payments will be $360(\$1,918.56) = \$690,681.60$, including payment of interest totaling to $370,681.60.

"Wow, that's a lot of interest money," exclaimed Glen. "The total is more than the loan amount."

"Indeed it is," responded Herkimer. "But remember that the lender has initially invested $320,000 and doesn't get that back for many years. The borrower is paying for the privilege of using that money." He asked the Pack to now consider a 20-year $250,000 mortgage at 7%. Handing transparencies to teams of two students, he asked them to

121

display relevant values along with the interest and principal portions for the first three payments.

It took a bit of time, but Frances and Janice came up with this handwritten display:

20 YR $250,000 LOAN AT 7%.

$$\text{PAYMENT} = \frac{(.07/12)(\$250,000)}{1-(1+.07/12)^{-240}}$$

$$= \$1,938.25.$$

ANALYSIS OF FIRST 3 PAYMENTS:

#	PYMT	INTEREST	PRINC.	PRINC. O/S
0	~~~	~~~	~~	$250,000
1	$1,938.25	$\left(\frac{.07}{12}\right)(\$250,000)$ $= \$1,458.33$	$\$1,938.25$ $-1,458.33$ $= \$479.92$	$\$250,000.00$ $-\$479.92$ $=\$249,520.08$
2	$1,938.25	$\left(\frac{.07}{12}\right)(\$249,547.08)$ $=\$1,455.53$	$\$1,938.25$ $-\$1,455.53$ $= \$482.72$	$\$249,520.08$ $- 482.72$ $=\$249,037.36$
3	$1,938.25	$\left(\frac{.07}{12}\right)(\$249,037.36)$ $= \$1,452.72$	$\$1,938.25$ $-\$1,452.72$ $= \$485.53$	$\$249,037.36$ $- \$485.53$ $=\$248,551.83$

"Nicely done," said Herkimer. "Note the large interest portions of the first payments. These interest amounts are tax deduction items so those with mortgages definitely want to take advantage of this fact. Now I'd like see which team can produce a spreadsheet

122

relating to a 6.5% mortgage of $500,000 displaying that a longer term mortgages involves smaller payments but considerably more paid interest over the term of the loan."

Roger and Stephen were first to respond:

	A	B	C	D	E
1	Spreadsheet by ROGER and STEPHEN				
2					
3	$500,000	<----Loan Amount			
4	0.065	<---Annual Rate (Compounded monthly)			
5					
6	Loan Term				
7	Years	Payment	Total Payments	Total Interest	
8	5	$9,783.07	$586,984.45	$86,984.45	
9	10	$5,677.40	$681,287.86	$181,287.86	
10	15	$4,355.54	$783,996.63	$283,996.63	
11	20	$3,727.87	$894,687.76	$394,687.76	
12	25	$3,376.04	$1,012,810.74	$512,810.74	
13	30	$3,160.34	$1,137,722.44	$637,722.44	
14	40	$2,927.28	$1,405,096.34	$905,096.34	
15					
16					

$500,000 Mortgage at 6.5%

$500,000 Mortgage at 6.5%

After praising the display by Stephen and Roger, Herkimer then put forth the challenge of producing a spreadsheet for a 30-year mortgage of $500,000 for interest rates

varying from 3% to 12%. After some serious computer
work, Carolyn and Wayne produced the following output:

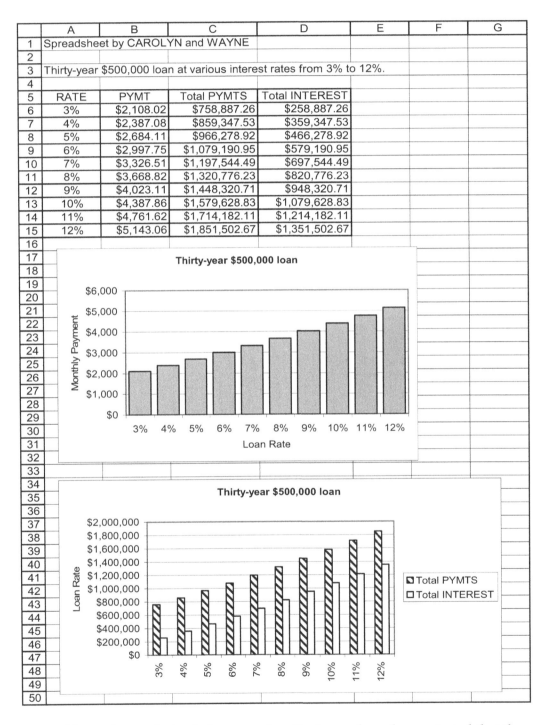

	A	B	C	D	E	F	G
1	Spreadsheet by CAROLYN and WAYNE						
2							
3	Thirty-year $500,000 loan at various interest rates from 3% to 12%.						
4							
5	RATE	PYMT	Total PYMTS	Total INTEREST			
6	3%	$2,108.02	$758,887.26	$258,887.26			
7	4%	$2,387.08	$859,347.53	$359,347.53			
8	5%	$2,684.11	$966,278.92	$466,278.92			
9	6%	$2,997.75	$1,079,190.95	$579,190.95			
10	7%	$3,326.51	$1,197,544.49	$697,544.49			
11	8%	$3,668.82	$1,320,776.23	$820,776.23			
12	9%	$4,023.11	$1,448,320.71	$948,320.71			
13	10%	$4,387.86	$1,579,628.83	$1,079,628.83			
14	11%	$4,761.62	$1,714,182.11	$1,214,182.11			
15	12%	$5,143.06	$1,851,502.67	$1,351,502.67			
16							
17							
18							
19							
20							
21							
22							
23							
24							
25							
26							
27							
28							
29							
30							
31							
32							
33							
34							
35							
36							
37							
38							
39							
40							
41							
42							
43							
44							
45							
46							
47							
48							
49							
50							

Herkimer always derived pleasure when Pack members demonstrated that they
understood the concepts he was attempting to teach.

On this day he was particularly happy. The Pack had taken note of two important facts relating to mortgages:

(1) Increasing the term for a mortgage means a smaller monthly payment. However, this increases the total amount of interest that will be paid on the loan.

(2) A small increase in the interest rate results in considerably more interest paid during the term of the loan.

ACTIVITY SET FOR SESSION #22

1. **ALCAL** activity:

(a) Use your calculator to check the following computation on the on the spreadsheet produced by Roger and Stephen: The monthly payment on a 20-year $500,000 loan at $i^{(12)} = 6.5\%$:

To be evaluated	$500000(.065/12)/[1-(1+.065/12)^{-240}]$
Calculator entry	500000*(.065/12)/(1-(1+.065/12)^240)
Number represented	

(b) Use your calculator to check the following computation on the on the spreadsheet produced by Carolyn and Wayne: The monthly payment on a 30-year $500,000 loan at $i^{(12)} = 7\%$:

To be evaluated	$500000(.07/12)/[1-(1+.07/12)^{-360}]$
Calculator entry	500000*(.07/12)/(1-(1+.07/12)^360)
Number represented	

2. REFLECTION & COMMENT activity:

As cosigner for a loan, you are being asked to guarantee the debt. You should consider this carefully before you do it. If the borrower does not pay the debt as agreed, you will have to pay it. You should be sure that yon can afford to pay if you have to, and that you want to accept this responsibility as a cosigner. As a cosigner for the loan, you may have to pay the full amount of the debt if the borrower does not pay. In addition, as cosigner you may also have to pay any late fees or collection costs.

(Peter Kenny, *The Danger of Cosigning a Loan*, www.articlebase.com, Jan 2008.)

3. The table below presents 10 mortgage descriptions. Fill in the missing portions:

Loan #	Loan Amount	Loan Term (Years)	Loan Rate	Monthly Payment	Total (All Payments)	Total Interest Paid
1	$800,000	30	6.40%			
2	$180,000	15	5.92%			
3	$600,000	20	6.18%			
4	$600,000	10	6.18%			
5	$350,000	20	4.80%			
6	$350,000	20	5.80%			
7	$350,000	20	6.80%			
8	$470,000	30	7.15%			
9	$470,000	20	7.15%			
10	$470,000	15	7.15%			

4. A $300,000 mortgage has a 6.2% rate. What will be the total interest paid over the term of the loan if the length of the term is (a) 30 years? (b) 20 years?

5. A $300,000 mortgage has a term of 30 years. What will be the total interest paid over the term of the loan if the rate is (a) 5.25%? (b) 6.25%?

6. Research this question: What is a **home equity loan**?

$$$
$$$

126

MORTGAGES - COMPLETE PAYMENT SCHEDULES

> *Economists report that a college education adds many thousands of dollars to a man's lifetime income - which he then spends sending his son to college.*
> -Bill Vaughn

The Pack realized that if one were to print out the entire payment schedule for a 30-year loan, one would have 360 lines of payment figures. They also realized that the interest portion of each payment would decrease as the years progressed and that the initial payments would be mostly interest. And, as Herkimer had previously indicated, the interest portions of the payments represented a tax-deductible item for homeowners paying off a mortgage. The total interest payments for a taxable year represented a considerable sum, particularly in the early years of a mortgage. The fact that this sum could be used to reduce one's tax liability was an encouragement of sorts for investing in home ownership.

Herkimer now put forth a major challenge. He wanted the Pack to construct a spreadsheet that would allow a user to input the mortgage amount, the loan interest rate, and the term (number of years) of the loan. The spreadsheet would then yield the entire loan payment schedule along with appropriate totals after the last payment.

Needless to say, this represented a challenge for this group of bright young students. However, while initially working in pairs, they combined their resources and ideas to accomplish the task. It took a good amount of time and the group had to work through logical errors and formatting errors, but they got the job done. The spreadsheet took up numerous pages (depending upon loan length, of course). Herkimer then asked the Pack to consider a $300,000 mortgage at 6%. He wanted a partial one-page schedule for a 30 year mortgage, a 20 year mortgage, and a 15 year mortgage. He asked the Pack to use the HIDE ROWS feature of spreadsheets and show the number analysis for just the first twelve payments, the middle-of-loan term twelve payments, and the final twelve payments along with the loan totals.

Another challenge! But the Pack was up to it. They weren't used to the HIDE ROWS feature, but using multiple computers and a considerable amount of idea exchanges, they figured out how to use it. The youngsters really felt a sense of accomplishment as they came up with the spreadsheets on the following pages:

NOTE TO READER: The complete 30-year mortgage spreadsheet appears in Appendix C

	A	B	C	D	E	F	G	H	I
1	MORTGAGE SPREADSHEET constructed by Herkimer's Stat Pack								
2	User inputs LOAN AMOUNT, INTEREST RATE, and NUMBER OF YEARS (TERM OF MORTGAGE).								
3									
4	$300,000	<---INPUT LOAN AMOUNT							
5	6.00%	<---INPUT INTEREST RATE (Annual Rate Compounded Monthly)							
6	30	<---INPUT NUMBER OF YEARS (TERM OF MORTGAGE)							
7									
8	$1,798.65	<----MONTHLY PAYMENT (Calculated)							
9									
10	Month #	Payment	Interest	Principal	Princ. Outstand.		% Interest in payment		
11	0				$300,000.00				
12	1	$1,798.65	$1,500.00	$298.65	$299,701.35		83.40%		
13	2	$1,798.65	$1,498.51	$300.14	$299,401.20		83.31%		
14	3	$1,798.65	$1,497.01	$301.65	$299,099.56		83.23%		
15	4	$1,798.65	$1,495.50	$303.15	$298,796.40		83.15%		
16	5	$1,798.65	$1,493.98	$304.67	$298,491.73		83.06%		
17	6	$1,798.65	$1,492.46	$306.19	$298,185.54		82.98%		
18	7	$1,798.65	$1,490.93	$307.72	$297,877.82		82.89%		
19	8	$1,798.65	$1,489.39	$309.26	$297,568.56		82.81%		
20	9	$1,798.65	$1,487.84	$310.81	$297,257.75		82.72%		
21	10	$1,798.65	$1,486.29	$312.36	$296,945.38		82.63%		
22	11	$1,798.65	$1,484.73	$313.92	$296,631.46		82.55%		
23	12	$1,798.65	$1,483.16	$315.49	$296,315.96		82.46%	1 year	
179									
180	169	$1,798.65	$1,108.31	$690.34	$220,971.93		61.62%		
181	170	$1,798.65	$1,104.86	$693.79	$220,278.14		61.43%		
182	171	$1,798.65	$1,101.39	$697.26	$219,580.88		61.23%		
183	172	$1,798.65	$1,097.90	$700.75	$218,880.13		61.04%		
184	173	$1,798.65	$1,094.40	$704.25	$218,175.88		60.85%		
185	174	$1,798.65	$1,090.88	$707.77	$217,468.10		60.65%		
186	175	$1,798.65	$1,087.34	$711.31	$216,756.79		60.45%		
187	176	$1,798.65	$1,083.78	$714.87	$216,041.93		60.26%		
188	177	$1,798.65	$1,080.21	$718.44	$215,323.48		60.06%		
189	178	$1,798.65	$1,076.62	$722.03	$214,601.45		59.86%		
190	179	$1,798.65	$1,073.01	$725.64	$213,875.81		59.66%		
191	180	$1,798.65	$1,069.38	$729.27	$213,146.53		59.45%	15 years	
359									
360	349	$1,798.65	$104.49	$1,694.16	$19,204.25		5.81%		
361	350	$1,798.65	$96.02	$1,702.63	$17,501.62		5.34%		
362	351	$1,798.65	$87.51	$1,711.14	$15,790.48		4.87%		
363	352	$1,798.65	$78.95	$1,719.70	$14,070.78		4.39%		
364	353	$1,798.65	$70.35	$1,728.30	$12,342.48		3.91%		
365	354	$1,798.65	$61.71	$1,736.94	$10,605.54		3.43%		
366	355	$1,798.65	$53.03	$1,745.62	$8,859.92		2.95%		
367	356	$1,798.65	$44.30	$1,754.35	$7,105.57		2.46%		
368	357	$1,798.65	$35.53	$1,763.12	$5,342.44		1.98%		
369	358	$1,798.65	$26.71	$1,771.94	$3,570.50		1.49%		
370	359	$1,798.65	$17.85	$1,780.80	$1,789.70		0.99%		
371	360	$1,798.65	$8.95	$1,789.70	$0.00		0.50%	30 years	
372									
373	TOTALS	$647,514.57	$347,514.57	$300,000.00	<---THESE ARE		30	YEAR TOTAL	

128

	A	B	C	D	E	F	G	H	I
1	MORGAGE SPREADSHEET…constructed by Herkimer's Stat Pack								
2	User inputs LOAN AMOUNT, INTEREST RATE, and NUMBER OF YEARS (TERM OF MORTGAGE).								
3									
4	$300,000	<---INPUT LOAN AMOUNT							
5	6.00%	<---INPUT INTEREST RATE (Annual Rate Compounded Monthly)							
6	20	<---INPUT NUMBER OF YEARS (TERM OF MORTGAGE)							
7									
8	$2,149.29	<----MONTHLY PAYMENT (Calculated)							
9									
10	Month #	Payment	Interest	Principal	Princ. Outstand.		% Interest in payment		
11	0				$300,000.00				
12	1	$2,149.29	$1,500.00	$649.29	$299,350.71		69.79%		
13	2	$2,149.29	$1,496.75	$652.54	$298,698.17		69.64%		
14	3	$2,149.29	$1,493.49	$655.80	$298,042.36		69.49%		
15	4	$2,149.29	$1,490.21	$659.08	$297,383.28		69.33%		
16	5	$2,149.29	$1,486.92	$662.38	$296,720.91		69.18%		
17	6	$2,149.29	$1,483.60	$665.69	$296,055.22		69.03%		
18	7	$2,149.29	$1,480.28	$669.02	$295,386.20		68.87%		
19	8	$2,149.29	$1,476.93	$672.36	$294,713.84		68.72%		
20	9	$2,149.29	$1,473.57	$675.72	$294,038.11		68.56%		
21	10	$2,149.29	$1,470.19	$679.10	$293,359.01		68.40%		
22	11	$2,149.29	$1,466.80	$682.50	$292,676.51		68.25%		
23	12	$2,149.29	$1,463.38	$685.91	$291,990.60		68.09%	1 year	
119									
120	109	$2,149.29	$1,036.60	$1,112.69	$206,207.26		48.23%		
121	110	$2,149.29	$1,031.04	$1,118.26	$205,089.01		47.97%		
122	111	$2,149.29	$1,025.45	$1,123.85	$203,965.16		47.71%		
123	112	$2,149.29	$1,019.83	$1,129.47	$202,835.69		47.45%		
124	113	$2,149.29	$1,014.18	$1,135.11	$201,700.58		47.19%		
125	114	$2,149.29	$1,008.50	$1,140.79	$200,559.79		46.92%		
126	115	$2,149.29	$1,002.80	$1,146.49	$199,413.29		46.66%		
127	116	$2,149.29	$997.07	$1,152.23	$198,261.07		46.39%		
128	117	$2,149.29	$991.31	$1,157.99	$197,103.08		46.12%		
129	118	$2,149.29	$985.52	$1,163.78	$195,939.30		45.85%		
130	119	$2,149.29	$979.70	$1,169.60	$194,769.70		45.58%		
131	120	$2,149.29	$973.85	$1,175.44	$193,594.26		45.31%	10 years	
239									
240	229	$2,149.29	$124.86	$2,024.43	$22,948.06		5.81%		
241	230	$2,149.29	$114.74	$2,034.55	$20,913.51		5.34%		
242	231	$2,149.29	$104.57	$2,044.73	$18,868.78		4.87%		
243	232	$2,149.29	$94.34	$2,054.95	$16,813.83		4.39%		
244	233	$2,149.29	$84.07	$2,065.22	$14,748.61		3.91%		
245	234	$2,149.29	$73.74	$2,075.55	$12,673.06		3.43%		
246	235	$2,149.29	$63.37	$2,085.93	$10,587.13		2.95%		
247	236	$2,149.29	$52.94	$2,096.36	$8,490.77		2.46%		
248	237	$2,149.29	$42.45	$2,106.84	$6,383.93		1.98%		
249	238	$2,149.29	$31.92	$2,117.37	$4,266.56		1.49%		
250	239	$2,149.29	$21.33	$2,127.96	$2,138.60		0.99%		
251	240	$2,149.29	$10.69	$2,138.60	$0.00		0.50%	20 years	
372									
373	TOTALS	$515,830.36	$215,830.36	$300,000.00	<---THESE ARE		20	YEAR TOTALS	

129

	A	B	C	D	E	F	G	H	I
1	MORGAGE SPREADSHEET...constructed by Herkimer's Stat Pack								
2	User inputs LOAN AMOUNT, INTEREST RATE, and NUMBER OF YEARS (TERM OF MORTGAGE).								
3									
4	$300,000	<---INPUT LOAN AMOUNT							
5	6.00%	<---INPUT INTEREST RATE (Annual Rate Compounded Monthly)							
6	15	<---INPUT NUMBER OF YEARS (TERM OF MORTGAGE)							
7									
8	$2,531.57	<----MONTHLY PAYMENT (Calculated)							
9									
10	Month #	Payment	Interest	Principal	Princ. Outstand.		% Interest in payment		
11	0				$300,000.00				
12	1	$2,531.57	$1,500.00	$1,031.57	$298,968.43		59.25%		
13	2	$2,531.57	$1,494.84	$1,036.73	$297,931.70		59.05%		
14	3	$2,531.57	$1,489.66	$1,041.91	$296,889.79		58.84%		
15	4	$2,531.57	$1,484.45	$1,047.12	$295,842.67		58.64%		
16	5	$2,531.57	$1,479.21	$1,052.36	$294,790.31		58.43%		
17	6	$2,531.57	$1,473.95	$1,057.62	$293,732.69		58.22%		
18	7	$2,531.57	$1,468.66	$1,062.91	$292,669.78		58.01%		
19	8	$2,531.57	$1,463.35	$1,068.22	$291,601.56		57.80%		
20	9	$2,531.57	$1,458.01	$1,073.56	$290,528.00		57.59%		
21	10	$2,531.57	$1,452.64	$1,078.93	$289,449.07		57.38%		
22	11	$2,531.57	$1,447.25	$1,084.33	$288,364.74		57.17%		
23	12	$2,531.57	$1,441.82	$1,089.75	$287,275.00		56.95%	1 year	
95									
96	85	$2,531.57	$963.20	$1,568.37	$191,072.04		38.05%		
97	86	$2,531.57	$955.36	$1,576.21	$189,495.83		37.74%		
98	87	$2,531.57	$947.48	$1,584.09	$187,911.74		37.43%		
99	88	$2,531.57	$939.56	$1,592.01	$186,319.73		37.11%		
100	89	$2,531.57	$931.60	$1,599.97	$184,719.75		36.80%		
101	90	$2,531.57	$923.60	$1,607.97	$183,111.78		36.48%		
102	91	$2,531.57	$915.56	$1,616.01	$181,495.77		36.17%		
103	92	$2,531.57	$907.48	$1,624.09	$179,871.68		35.85%		
104	93	$2,531.57	$899.36	$1,632.21	$178,239.47		35.53%		
105	94	$2,531.57	$891.20	$1,640.37	$176,599.09		35.20%		
106	95	$2,531.57	$883.00	$1,648.58	$174,950.52		34.88%		
107	96	$2,531.57	$874.75	$1,656.82	$173,293.70		34.55%	8 years	
108									
180	169	$2,531.57	$147.07	$2,384.50	$27,029.65		5.81%		
181	170	$2,531.57	$135.15	$2,396.42	$24,633.22		5.34%		
182	171	$2,531.57	$123.17	$2,408.40	$22,224.82		4.87%		
183	172	$2,531.57	$111.12	$2,420.45	$19,804.37		4.39%		
184	173	$2,531.57	$99.02	$2,432.55	$17,371.82		3.91%		
185	174	$2,531.57	$86.86	$2,444.71	$14,927.11		3.43%		
186	175	$2,531.57	$74.64	$2,456.93	$12,470.18		2.95%		
187	176	$2,531.57	$62.35	$2,469.22	$10,000.96		2.46%		
188	177	$2,531.57	$50.00	$2,481.57	$7,519.39		1.98%		
189	178	$2,531.57	$37.60	$2,493.97	$5,025.42		1.49%		
190	179	$2,531.57	$25.13	$2,506.44	$2,518.98		0.99%		
191	180	$2,531.57	$12.59	$2,518.98	$0.00		0.50%	15 years	
372									
373	TOTALS	$455,682.69	$155,682.69	$300,000.00	<---THESE ARE		15	YEAR TOTALS	

"Absolutely fantastic," concluded Herkimer.

ACTIVITY SET FOR SESSION #23

1. **ALCAL** activity:

   (a) Use your calculator to check the total of all the payments on the Stat Pack spreadsheet relating to the 30-year $300,000 mortgage at 6%:

To be evaluated	$360(300000)(.06/12)/[1-(1+.06/12)^{-360}]$
Calculator entry	360*300000*(.06/12)/(1-(1+.06/12)^360)
Number represented	

   (b) Use your calculator to check the total of all the payments on the Stat Pack spreadsheet relating to the 15-year $300,000 mortgage at 6%:

To be evaluated	$180(300000)(.06/12)/[1-(1+.06/12)^{-180}]$
Calculator entry	180*300000*(.06/12)/(1-(1+.06/12)^180)
Number represented	

2. REFLECTION & COMMENT activity:

   Essentially, using a home equity loan represents some choice - to use an asset to finance a new purchase, and to use a second mortgage to access that asset as opposed to other methods of financing. Home equity loans carry the same obligations as your primary mortgage. Specifically, both primary and secondary mortgages are loans made against the value of a property, with the property subject to foreclosure if the loan obligations are not met. So, this is serious business - a $10,000 home equity loan can put your $300,000 home in play if you do not meet your obligation to make principal and interest payments on schedule.

   (Richard Barrington, www.debthelp.com, 2009)

3. For those who can work the mathematics of spreadsheets, construct a spreadsheet that will display a complete mortgage payment schedule when a user supplied the mortgage amount, the interest rate, and the term of the mortgage. [A reminder that Appendix C does have a complete payment schedule for a 30-year mortgage.]

4. Use the mortgage schedules in the vignette to complete the following table relating to a $300,000 mortgage at 6%. In doing so, contrast the differences that the term of the loan makes, and remember that interest payments are tax deductible items:

	15 year-mortgage	20-year mortgage	30-year mortgage
Monthly payment			
Total all payments			
Total interest paid			
Total interest in first 12 payments			
Total interest in middle 12 payments			
Total interest in final 12 payments			

5. Consider a $400,000 thirty-year mortgage. Fill in the table below for the indicated loan rates:

	5%	6%	7%	8%
Monthly payment				
Total all payments				
Total interest paid				

6. Consider a $400,000 fifteen-year mortgage. Fill in the table below for the indicated loan rates. Contrast the results with those obtained in activity #3:

	5%	6%	7%	8%
Monthly payment				
Total all payments				
Total interest paid				

$$
$$$

Session #24
THE CREDIT CARD TRAP

Someone stole all of my credit cards, but I won't be reporting it. The thief spends less than my wife did.
-Henny Youngman

Pack members were aware that credit card debt was a big issue for many people. They didn't know exactly why this type of problem was so widespread, but they were about to find out.

"It's primarily a problem of good folks not understanding what develops when one makes minimal payments on monthly credit card bills," said Herkimer. He went on to explain that credit card companies shouldn't be blamed for this nationwide problem. "If one uses credit cards wisely, they represent a very convenient payment method since you don't have to carry large amounts of cash. And, if you pay the total amount you owe each month you have the equivalent of an interest-free loan for the month. This takes some discipline, of course, since you must be sure that you don't charge more than you can afford to pay in one lump sum. Interest rates on credit cards are high, usually more than 20%. That's why it is important to understand both the good and bad features relating to them."

Herkimer went on to explain that many American families have thousands of dollars of credit card debt. They get in so deep that they have tremendous problems "digging themselves out" of this type of debt. Herkimer wanted to demonstrate how this type of situation can develop. He presented a situation in which he used a credit card to purchase a fancy $1,400 television set. Then, as expected, the first credit card billing came a few weeks later containing this information:

> **Purchases and Debits: $1,400.00**
> **Total Minimum Due: $30.00**
> **Annual Percentage Rate: 24%***
> **Effective Annual Percentage Rate: 26.82%**
> *** Rate <u>compounded monthly</u>**

The Pack quickly used previously knowledge to calculate $(1 + 0.24/12)^{12} - 1 = 0.2682$.

"Now," Herkimer continued, "if I paid the entire $1,400, then I have had an interest-free loan for that amount for about a month. But let's assume that I am simply delighted to have that wonderful TV set for just $30 a month. Going to the chalkboard, he outlined the structure for the first few payments.

133

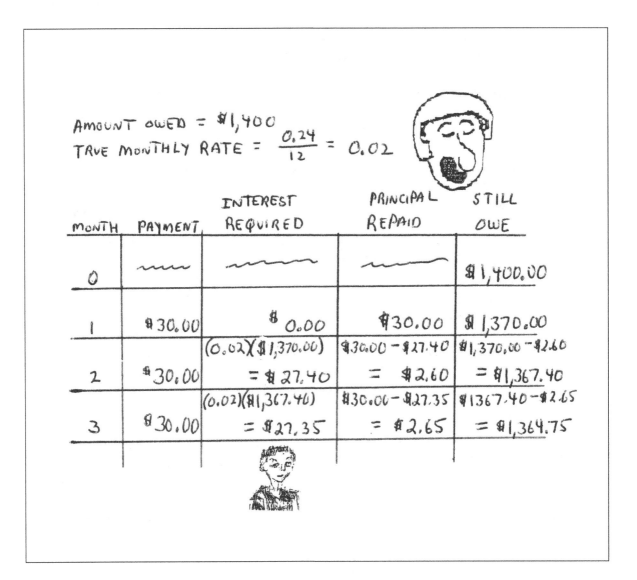

AMOUNT OWED = $1,400
TRUE MONTHLY RATE = $\frac{0.24}{12}$ = 0.02

MONTH	PAYMENT	INTEREST REQUIRED	PRINCIPAL REPAID	STILL OWE
0	~~~	~~~	~~~	$1,400.00
1	$30.00	$0.00	$30.00	$1,370.00
2	$30.00	(0.02)($1,370.00) = $27.40	$30.00 − $27.40 = $2.60	$1,370.00 − $2.60 = $1,367.40
3	$30.00	(0.02)($1,367.40) = $27.35	$30.00 − $27.35 = $2.65	$1367.40 − $2.65 = $1,364.75

"Oh, wow!" responded an alert Stephen, "I think I see where this is going. It's going to take forever to pay off the debt with just $30 a month. Most of the payments are total interest."

"Yes," said Herkimer. "Now realize you can pay more than the $30 any month, but let's try to get a look at what happens if you just pay the minimum amount each month. How about a nice spreadsheet that will show this for us over, let's say, a period of 4 years. This shouldn't be too much of a challenge at this stage of the game since it is somewhat similar to mortgage spreadsheets.

The team of Brenda and Glen came up with the following interactive spreadsheet. One could actually input the amount owed, the interest rate, and the minimum payment required. The sheet would adjust to the values inputted.

	A	B	C	D	E	F
1	Spreadsheet by BRENDA and GLEN					
2	Credit Card Example					
3	Amount owed			$1,400.00		
4	Yearly rate(nominal)			0.24		
5	Monthly rate			0.02		
6	Minimum monthly payment			$30.00		
7						
8	Month	Payment	Interest	Princ. Repaid	Still owe	
9	0				$1,400.00	
10	1	$30.00	$0.00	$30.00	$1,370.00	
11	2	$30.00	$27.40	$2.60	$1,367.40	
12	3	$30.00	$27.35	$2.65	$1,364.75	
13	4	$30.00	$27.29	$2.71	$1,362.04	
14	5	$30.00	$27.24	$2.76	$1,359.28	
15	6	$30.00	$27.19	$2.81	$1,356.47	
16	7	$30.00	$27.13	$2.87	$1,353.60	
17	8	$30.00	$27.07	$2.93	$1,350.67	
18	9	$30.00	$27.01	$2.99	$1,347.68	
19	10	$30.00	$26.95	$3.05	$1,344.64	
20	11	$30.00	$26.89	$3.11	$1,341.53	
21	12	$30.00	$26.83	$3.17	$1,338.36	
22	13	$30.00	$26.77	$3.23	$1,335.13	
23	14	$30.00	$26.70	$3.30	$1,331.83	
24	15	$30.00	$26.64	$3.36	$1,328.47	
25	16	$30.00	$26.57	$3.43	$1,325.04	
26	17	$30.00	$26.50	$3.50	$1,321.54	
27	18	$30.00	$26.43	$3.57	$1,317.97	
28	19	$30.00	$26.36	$3.64	$1,314.33	
29	20	$30.00	$26.29	$3.71	$1,310.61	
30	21	$30.00	$26.21	$3.79	$1,306.83	
31	22	$30.00	$26.14	$3.86	$1,302.96	
32	23	$30.00	$26.06	$3.94	$1,299.02	
33	24	$30.00	$25.98	$4.02	$1,295.00	
34	25	$30.00	$25.90	$4.10	$1,290.90	
35	26	$30.00	$25.82	$4.18	$1,286.72	
36	27	$30.00	$25.73	$4.27	$1,282.46	
37	28	$30.00	$25.65	$4.35	$1,278.10	
38	29	$30.00	$25.56	$4.44	$1,273.67	
39	30	$30.00	$25.47	$4.53	$1,269.14	
40	31	$30.00	$25.38	$4.62	$1,264.52	
41	32	$30.00	$25.29	$4.71	$1,259.81	
42	33	$30.00	$25.20	$4.80	$1,255.01	
43	34	$30.00	$25.10	$4.90	$1,250.11	
44	35	$30.00	$25.00	$5.00	$1,245.11	
45	36	$30.00	$24.90	$5.10	$1,240.01	
46	37	$30.00	$24.80	$5.20	$1,234.81	
47	38	$30.00	$24.70	$5.30	$1,229.51	
48	39	$30.00	$24.59	$5.41	$1,224.10	
49	40	$30.00	$24.48	$5.52	$1,218.58	
50	41	$30.00	$24.37	$5.63	$1,212.95	
51	42	$30.00	$24.26	$5.74	$1,207.21	
52	43	$30.00	$24.14	$5.86	$1,201.36	
53	44	$30.00	$24.03	$5.97	$1,195.39	
54	45	$30.00	$23.91	$6.09	$1,189.29	
55	46	$30.00	$23.79	$6.21	$1,183.08	
56	47	$30.00	$23.66	$6.34	$1,176.74	
57	48	$30.00	$23.53	$6.47	$1,170.28	<--still owe
58						
59	Totals	$1,440.00	$1,210.28	$229.72		

135

"Holy Herkimer," exclaimed Roger, "after four years only about $230 of the bill has been paid off and about $1,170 is still owed. How long would it take to get this thing paid off?"

Brenda and Glen actually extended the table and hid most of the middle rows. Their spreadsheet (below) indicated that after 124 months the amount owned would be $14.87. The final billing would be for $14.87 + (0.02)($14.87) = $15.17. The Pack was somewhat astounded to see that if one simply made the minimum payment required it would take over 10 years to pay off the $1,400 debt. And, one would pay a total of $3,735.17 including $2,335.17 in interest.

	A	B	C	D	E	F
1	Spreadsheet by BRENDA and GLEN					
2	Credit Card Example					
3	Amount owed			$1,400.00		
4	Yearly rate(nominal)			0.24		
5	Monthly rate			0.02		
6	Minimum monthly payment			$30.00		
7						
8	Month	Payment	Interest	Princ. Repaid	Still owe	
9	0				$1,400.00	
10	1	$30.00	$0.00	$30.00	$1,370.00	
11	2	$30.00	$27.40	$2.60	$1,367.40	
12	3	$30.00	$27.35	$2.65	$1,364.75	
13	4	$30.00	$27.29	$2.71	$1,362.04	
14	5	$30.00	$27.24	$2.76	$1,359.28	
129						Hidden rows
130	121	$30.00	$2.56	$27.44	$100.53	
131	122	$30.00	$2.01	$27.99	$72.54	
132	123	$30.00	$1.45	$28.55	$43.99	
133	124	$30.00	$0.88	$29.12	$14.87	
134	125	$15.17	$0.30	$14.87	$0.00	
135						
136	Totals	$3,735.17	$2,335.17	$1,400.00		

Note to Reader: A complete schedule appears in Appendix C

Herkimer stressed once again that one wouldn't have to make just the minimum required payment each month. One wouldn't even have to make equal payments, just as long as the payment was at least the minimum required.

"What would happen if one paid $100 each month?" asked Herkimer. He indicated that the spreadsheet by Brenda and Glen could be used to check this out. When the *still owe* column started producing negative numbers, the debt would be paid off. (On a spreadsheet, a negative number appears in parentheses.) The Pack used the spreadsheet with a $100 monthly payment and observed the following:

	A	B	C	D	E
1	Spreadsheet by BRENDA and GLEN				
2	Credit Card Example				
3	Amount owed			$1,400.00	
4	Yearly rate(nominal)			0.24	
5	Monthly rate			0.02	
6	Minimum monthly payment			$100.00	
7					
8	Month	Payment	Interest	Princ. Repaid	Still owe
9	0				$1,400.00
10	1	$100.00	$0.00	$100.00	$1,300.00
11	2	$100.00	$26.00	$74.00	$1,226.00
12	3	$100.00	$24.52	$75.48	$1,150.52
13	4	$100.00	$23.01	$76.99	$1,073.53
14	5	$100.00	$21.47	$78.53	$995.00
15	6	$100.00	$19.90	$80.10	$914.90
16	7	$100.00	$18.30	$81.70	$833.20
17	8	$100.00	$16.66	$83.34	$749.86
18	9	$100.00	$15.00	$85.00	$664.86
19	10	$100.00	$13.30	$86.70	$578.16
20	11	$100.00	$11.56	$88.44	$489.72
21	12	$100.00	$9.79	$90.21	$399.52
22	13	$100.00	$7.99	$92.01	$307.51
23	14	$100.00	$6.15	$93.85	$213.66
24	15	$100.00	$4.27	$95.73	$117.93
25	16	$100.00	$2.36	$97.64	$20.29
26	17	$100.00	$0.41	$99.59	($79.31)

Herkimer explained that only $20.29 was still owed after the 16th payment. The credit card bill for the 17th month would include $0.41 in interest and hence request a total of $20.70. The payment of $100 would not be required. The Pack concluded that with $100 payments the debt of $1,400 would be paid of within 1 1/2 years with an interest total of $220.69.

Herkimer went on to explain that credit card companies are businesses and need to make money to survive. "In the above example, the credit card company pays the $1,400 to the seller so that I can have that nice TV set. In essence, they lend me the money and I have to pay off the loan to them. Credit card companies make short term loans for relatively small amounts when compared to a home mortgage, for instance. They require no background check and it is very easy to get a credit card. They charge high interest rates, but that's to be expected considering the risk they take in making the loans. Since the loans are so easy to get, you must be careful since it's not difficult to get carried away with the purchasing power they offer. It's not a something for nothing situation. You eventually have to pay off what you borrowed."

The Pack realized they had just completed one very valuable and important lesson in finance.

ACTIVITY SET FOR SESSION #24

1. **ALCAL** activity: The request references the spreadsheet constructed by Brenda and Glen relating to minimum payments of $30 on a $1,400 credit card debt with an interest rate of $i^{(12)} = 24\%$. Use your calculator to check the "still owe" total of all after the second payment of $30 is made:

To be evaluated	1370-[30-(.24/12)1370]
Calculator entry	1370-(30-(.24/12)*1370)
Number represented	

2. REFLECTION & COMMENT activity:

Say you paid $19 for a pepperoni pizza by charging it to your credit card. If you carry no balance on your credit card and pay your bill within the 20-to-25-day grace period, the pizza won't cost an extra dime. If you simply add the cost of the pizza as a topping to all your other revolving debt, that "inexpensive" dinner out would have cost a total of $40.04. Did you want to pay that much for a pizza? Probably not. Next time, when carrying balances, pay cash.

("The New Rules of Borrowing," *Consumer Reports,* July 1999)

In problems 3-5, complete the tables relating to the first three payments of a $5,000 credit card debt with a 20% annual rate compounded monthly. Minimum required monthly payment = $90.

3.

Month	Payment	Interest Required	Principal Repaid	Still Owe
0				$5,000.00
1	$90.00			
2	$90.00			
3	$90.00			

138

4.

Month	Payment	Interest Required	Principal Repaid	Still Owe
0				$5,000.00
1	$200.00			
2	$600.00			
3	$100.00			

5.

Month	Payment	Interest Required	Principal Repaid	Still Owe
0				$5,000.00
1	$90.00			
2	$1,000.00			
3	$800.00			

6. Assume your credit card interest rate is 18.25% and that you use your credit card to purchase a $3,200 item. For each of the first three months after the purchase you pay $1,000. Construct a payment schedule that has you paying off the debt with the 4th payment. Remember that the final payment must include the required interest.

$$
$$$

Session #25
CREDIT - A DANGEROUS WAY TO LIVE BEYOND YOUR MEANS

> *The only reason I made a commercial for American Express was to pay for my American Express bill.*
> -Peter Ustinov

Herkimer wanted to be sure that the young Pack members understood both the conveniences and the problems associated with credit card use. Discussion involved the fact that credit cards make it possible for one to avoid carrying around large amounts of cash. On trips, for instance, they can be used to pay for meals, hotels, rental cars and other costly items. They can also allow the purchase of luxury items such as television sets or fancy furniture. "The key," said Herkimer, "is to not charge more than you can afford to pay off in the very near future. Herein lies the real problem for many good folks. They start to live beyond their means by not immediately paying cash for purchased items. Eventually, financial reality catches up with them when they realize the debt they have actually incurred. The idea of paying small amounts over time for desired items is appealing, but far too few people understand that the early small payments are almost all interest and that the principal owned is being reduced only by a very small amount with each payment. You have seen this in the examples we illustrated in an earlier session."

Herkimer and the Pack spent considerable time discussing credit cards and using the Internet to find out interesting statistics related to credit card use. Some of the things they found:

While statistical data varies by source, average credit card debt for cardholders is somewhere around $8,000.

Some people have over $20,000 in credit card debt.

Most credit cards have limits. For instance, a card with a $5,000 limit would not allow the user to charge more than $5,000 with the card. When the user reached the $5,000 limit the card would not be accepted for purchases. The card would be *maxed out*.

Credit cards are easy to obtain and many people have numerous credit cards. While the cards have limits, when one is *maxed out,* another card is used for the purchase. Many maxed out credit cards means considerable credit card debt. And, making the minimum payment on numerous credit card bills is a sure road to financial difficulty.

A credit card rate is frequently an annual rate compounded monthly. In some cases it might be a rate that is compounded daily. In any case, if you assume it is a rate compounded monthly, you will be able to get a realistic picture of what happens if you just make small payments.

Credit card rates can vary considerably. Introductory rates are often low, but these quickly change to higher rates after a specified period of time. Credit card rates are generally between 14% and 28%, with the higher rates usually associated with cash

advances. **(You can get cash using your credit card, but the amount you can borrow is usually less than the credit card limit.)**

Mortgage interest is a tax deductible item, credit card interest is not.

Herkimer went on to use the overhead to illustrate that even relatively small differences in payments and interest rates can result in very large long term differences in total payments and interest paid. Using previously-constructed Pack spreadsheets, he displayed a schedule for a debt of $8,000 at a rate of 18%. As the sheet indicates, with a minimum monthly payment of $140 it would take 131 months (almost 11 years) to pay off the debt. Total payments would be $18,297.92 with total interest payment of $10,297.92.

	A	B	C	D	E	F
3	$8,000.00	Credit card debt				
4	18%	Credit card annual rate				
5	0.015	True monthly rate.				
6	$140.00	Minimum monthly payment				
7						
8	Month #	Payment	Interest	Principal	Still Owe	
9	0				$8,000.00	
10	1	$140.00	$120.00	$20.00	$7,980.00	
11	2	$140.00	$119.70	$20.30	$7,959.70	
12	3	$140.00	$119.40	$20.60	$7,939.10	
13	4	$140.00	$119.09	$20.91	$7,918.18	
14	5	$140.00	$118.77	$21.23	$7,896.95	
15	6	$140.00	$118.45	$21.55	$7,875.41	
16	7	$140.00	$118.13	$21.87	$7,853.54	
17	8	$140.00	$117.80	$22.20	$7,831.34	
18	9	$140.00	$117.47	$22.53	$7,808.81	
19	10	$140.00	$117.13	$22.87	$7,785.95	
20	11	$140.00	$116.79	$23.21	$7,762.73	
21	12	$140.00	$116.44	$23.56	$7,739.18	
22	13	$140.00	$116.09	$23.91	$7,715.26	Hidden rows
129	120	$140.00	$22.38	$117.62	$1,374.24	
130	121	$140.00	$20.61	$119.39	$1,254.85	
131	122	$140.00	$18.82	$121.18	$1,133.67	
132	123	$140.00	$17.01	$122.99	$1,010.68	
133	124	$140.00	$15.16	$124.84	$885.84	
134	125	$140.00	$13.29	$126.71	$759.13	
135	126	$140.00	$11.39	$128.61	$630.51	
136	127	$140.00	$9.46	$130.54	$499.97	
137	128	$140.00	$7.50	$132.50	$367.47	
138	129	$140.00	$5.51	$134.49	$232.98	
139	130	$140.00	$3.49	$136.51	$96.48	
140	131	$97.92	$1.45	$96.47	$0.00	
141						
142	Totals	$18,297.92	$10,297.92	$8,000.00		

"Now let's see what happens with different monthly payments," said Herkimer. The Pack all had the spreadsheet with no hidden rows on their respective team computers. They discovered that …

> **If monthly payment = $160, debt is paid off in 94 months with total interest payments approximately = $6897.**

If monthly payment = $200, debt is paid off in 62 months with total interest payments approximately = $4300

If monthly payment = $500, debt is paid off in 18 months with total interest payments approximately = $1215.

"Wow, wow, wow!" chirped Janice, "the more you can pay monthly, the better off you are in the long run."

"Yup," responded Darren, "but don't forget that if you pay off the total at the first billing, you have an interest-free loan for about a month. That's what Herkimer told us. I see what he means."

Herkimer was delighted with these comments. "You are learning," he said. "Now let's use that spreadsheet to see how the interest rate affects the overall picture. Let's assume that you are paying off $8,000 in credit card debt and that you will make the minimum payment of $140 a month. You have the situation for a rate of 18%. Check it out for rates 16%, 17%, 19%, and 20%."

Pack members took to the spreadsheet once again, inserting the various interest rates. It wasn't long before the teams agreed on these figures:

Debt = $8,000, monthly payment = $140

Rate	# months to pay off debt	Total interest paid
16%	109	$7,169
17%	118	$8,503
18%	131	$10,297
19%	150	$12,550
20%	185	$15,170

Herkimer concluded, "It should be obvious that a small difference in interest rates can make a big difference in total interest payments. While it makes sense to try to find a credit card with a relatively low interest rate, this is important only if you don't plan to pay the total amount you owe each month. And you can pay off what you owe if you plan carefully and don't spend more than you can afford to pay. This takes planning and discipline, and the amount of credit card debt in the country certainly suggests that many good folks don't plan well and lack the will power needed to avoid overspending."

Pack members knew they were learning a very valuable lesson about money.

HERKIMER'S FASCINATING FINANCIAL FACTS:

The first type of credit card appeared in the 1910s as a department store card. It was limited to purchases made at a specific store. In 1950 the Diner's Club introduced the first credit card that could be used at multiple locations. The locations were restaurants, but it was the first time credit was extended beyond a specific location.

ACTIVITY SET FOR SESSION #25

1. **ALCAL** activity: The request references the spreadsheet illustrated by Herkimer during the last session relating to minimum payments of $140 on a $8,000 credit card debt with an interest rate of $i^{(12)} = 18\%$. Use your calculator to check the "still owe" total of all after the third payment of $1400 is made:

To be evaluated	7959.70-[140-(.18/12)7959.70]
Calculator entry	7959.70-(140-(.18/12)*7959.70)
Number represented	

2. REFLECTION & COMMENT activity:

   Lenders would prefer you didn't worry about how much a debt will cost or how long it will take you to repay. All they want you to think about is the minimum monthly payment that will keep your account current.

   ("The New Rules of Borrowing," *Consumer Reports*, July 1999)

3. Use the Internet and search engines to find information relating to credit card debt in the United States. Using quotes around phrases like **"credit card debt"** keys in on sites that relate to the topic between the quotes. Phrases that yield lots of interesting information include:

   "credit card tips"   **"introductory credit card rates"**   **"credit card traps"**

4. Examine some actual credit card bills. Note the amount owed and the minimum payment that is required. Produce a schedule to demonstrate the effects of paying only the minimum required amount each month. [Note: If the billing says that the rate is a daily rate of some sort, simply treat it as a monthly rate and this will allow you to produce a very good approximation to the actual situation that will develop by making small monthly payments.]

5. Find examples of credit card advertisements that offer introductory rates. Some companies may even offer a 0% rate for a brief period of time. These might be beneficial for those who use credit cards wisely, but many people don't anticipate the financial impact that will set in when the introductory period expires and regular rates are applied. Complete the following table to display 6 months of payments for a $5,000 credit card debt with an introductory rate of 0% for the first three months. After the introductory period, the rate becomes 18%. A minimum monthly payment of $75 is required:

Month	Payment	Interest	Principal Repaid	Still Owe
0				$5,000.00
1	$75.00	$0.00	$75.00	$4,925.00
2	$75.00	$0.00	$75.00	$4,850.00
3	$75.00	$0.00	$75.00	$4,775.00
4	$75.00			
5	$75.00			
6	$75.00			

$$
$$

144

HERKIMER'S QUICK QUIZ
SESSIONS 21-25.

Find the correct response for
each question in the 25 cells
at the bottom of the page.

All rates are annual rates compounded monthly.

1. The monthly payment on a 15% five-year $22,000 loan is _____

2. A 30-year mortgage of $320,000 at 6.8% has a monthly payment of _____

3. What is the total amount of interest paid on the loan described in activity #2?

4. Consider a 20-year $440,000 mortgage at 5.95%. What is the total of all of the monthly payments?

5. A credit card holder has been paying a required minimum of $70 a month. If the interest rate is 21% and the outstanding balance after the 18th payment is $3,186, what amount of the 19th payment will be interest?

6. A credit card with a 23.65% rate has an outstanding balance of $2,412. If a check for $2,000 is sent at the next billing, what amount in this payment represents principal?

7. If a $175,000 twenty year mortgage has a rate of 7.1%, what is the sum of all of the principal payments over the twenty year period?

8. For the mortgage described in activity #7, what is the total interest paid over the twenty year period?

9. Consider a $380,000 thirty year mortgage at 6.1%. What is the principal outstanding after the 12th installment payment?

10. For the mortgage described in activity #9, what is the principal outstanding after 29 years?

	A	B	C	D	E
1	$23,553	$61.34	$87.88	$167,500	$402,357
2	$798,326	$674.22	$2,086.16	$181,000	$25,552
3	$153,151	2,673.21	$369,875	$1,952.46	$55.76
4	$48.58	$175,000	$753,508	$523.38	$811,235
5	$1,912.34	$431,018	$26,742	$149,145	$375,420

Session #26
BASIC CONCEPT BEHIND RETIREMENT PLANS

> *Creditors have better memories than debtors.*
> -Benjamin Franklin

Retirement is a topic of interest to just about everybody. Most workers have some type of retirement plan that accumulates money that has been invested so that it will be available for use when actual retirement happens. In many cases, a worker will invest a small portion of his salary into a retirement plan each month and his employer will match that amount. The invested amounts earn interest over a period of many years. Herkimer reminded the Pack that they had witnessed the power of compounding in earlier sessions.

"Once withdrawals from the fund start, how long will retirement money last?" asked Herkimer. The Pack realized that the question was somewhat vague but they knew that Herkimer wanted them to think about a broad situation. Discussions within the group came to the conclusion that to respond to the question one would need to know (1) the amount in the fund when withdrawals began, (2) the amount of the withdrawals, and (3) the interest rate earned by the fund.

"OK," Herkimer began after the discussions, "let's assume that an individual has $500,000 in a retirement fund that earns a true annual rate of 6%. She wants to withdraw $35,000 from the fund each year. How long will she be able to do this? In other words, with this plan how long will the fund last? Think about this, and then let's have a team go to the chalkboard and show us how we would get started on this problem."

After team discussions and a bit of paperwork, Frances and Carolyn picked up some chalk and wrote the following:

YEAR	AMOUNT IN FUND	AMOUNT WITHDRAWN	AMOUNT REMAINING	ACCUM AT 6%
1	$500,000	$35,000	$465,000	$465,000(1.06) = $492,900
2	$492,900	$35,000	$457,900	$457,900(1.06) = $485,374
3	$485,374	$35,000	$450,374	$450,374(1.06) = $477,396

"We would simply continue this table until we got a negative figure in the last column," said Frances after Carolyn had done the writing. "A negative figure would mean the fund was out of money. A spreadsheet analysis would be useful here."

146

Spreadsheet construction was very familiar to the Pack now. Glen and Roger produced the following spreadsheet table, noting that the fund would run out of money in 29 years. It was here that the fund amount turned negative, indicating that a full payment of $35,000 could not be supplied with the remaining money.

	A	B	C	D	E
1	Spreadsheet by GLEN and ROGER.				
2					
3	$500,000	<---Initial fund amount			
4	6%	<--True annual interest rate earned by fund			
5	$35,000	<---Amount withdrawn each year			
6				Fund amount	Fund amount with
7	Year	Amount in fund	Amount withdrawn	remaining	accumulated interest
8	1	$500,000	$35,000	$465,000	$492,900
9	2	$492,900	$35,000	$457,900	$485,374
10	3	$485,374	$35,000	$450,374	$477,396
11	4	$477,396	$35,000	$442,396	$468,940
12	5	$468,940	$35,000	$433,940	$459,977
13	6	$459,977	$35,000	$424,977	$450,475
14	7	$450,475	$35,000	$415,475	$440,404
15	8	$440,404	$35,000	$405,404	$429,728
16	9	$429,728	$35,000	$394,728	$418,412
17	10	$418,412	$35,000	$383,412	$406,416
18	11	$406,416	$35,000	$371,416	$393,701
19	12	$393,701	$35,000	$358,701	$380,223
20	13	$380,223	$35,000	$345,223	$365,937
21	14	$365,937	$35,000	$330,937	$350,793
22	15	$350,793	$35,000	$315,793	$334,741
23	16	$334,741	$35,000	$299,741	$317,725
24	17	$317,725	$35,000	$282,725	$299,689
25	18	$299,689	$35,000	$264,689	$280,570
26	19	$280,570	$35,000	$245,570	$260,304
27	20	$260,304	$35,000	$225,304	$238,822
28	21	$238,822	$35,000	$203,822	$216,052
29	22	$216,052	$35,000	$181,052	$191,915
30	23	$191,915	$35,000	$156,915	$166,330
31	24	$166,330	$35,000	$131,330	$139,209
32	25	$139,209	$35,000	$104,209	$110,462
33	26	$110,462	$35,000	$75,462	$79,990
34	27	$79,990	$35,000	$44,990	$47,689
35	28	$47,689	$35,000	$12,689	$13,450
36	29	$13,450	$35,000	($21,550)	($22,843)
37	30	($22,843)	$35,000	($57,843)	($61,313)

It was noted that in 28 years, a total of (28)($35,000) =$980,000 was withdrawn and the remaining amount was $13,450. Herkimer also wanted the Pack to realize that the initial $500,000 was an accumulation of deposits and interest. The owner of the retirement fund "invested" considerably less than $500,000 of her own money. The power of compounding was once again being illustrated for the Pack.

All Pack teams had similar spreadsheets. Herkimer asked the teams to use them to find how long a $500,000 fund with $35,000 annual withdrawals would last at various annual fund rates. And, assuming a 5% annual fund rate, how long the fund would last with various annual withdrawals?

Pack teams tried a variety of financial scenarios on their spreadsheets. After a short period of time, Herkimer put the following transparency on the overhead and asked the Pack to check the figures using their respective spreadsheet programs.

Fund	Annual rate earned by fund	Annual amount withdrawn	Can withdraw annual amount for…	Total of annual payments withdrawn	Remaining in fund after last full withdrawal
$500,000	3%	$35,000	18 years	$630,000	$7,126
$500,000	4%	$35,000	20 year	$700,000	$11,640
$500,000	5%	$35,000	23 years	$805,000	$13,192
$500,000	6%	$35,000	28 years	$980,000	$13,450
$500,000	7%	$35,000	40 years	$1,400,000	$10,894
$500,000	5%	$30,000	32 years	$960,000	$10,558
$500,000	5%	$35,000	23 years	$805,000	$13,192
$500,000	5%	$40,000	18 years	$720,000	$21,749
$500,000	5%	$45,000	15 years	$675,000	$19,877
$500,000	5%	$50,000	13 years	$650,000	$12,283

There was general agreement that the displayed financial amounts were indeed correct.

Herkimer wanted the Pack to understand that it would be possible for a retirement fund to actually increase if the withdrawals were "recovered" by the interest rate. He used the spreadsheet constructed by Glen and Roger to illustrate this fact:

	A	B	C	D	E
1	Spreadsheet by GLEN and ROGER.				
2					
3	$500,000	<---Initial fund amount			
4	6%	<--True annual interest rate earned by fund			
5	$25,000	<---Amount withdrawn each year			
6				Fund amount remaining	Fund amount with accumulated interest
7	Year	Amount in fund	Amount withdrawn		
8	1	$500,000	$25,000	$475,000	$503,500
9	2	$503,500	$25,000	$478,500	$507,210
10	3	$507,210	$25,000	$482,210	$511,143

Herkimer asked the Pack to notice that if the annual withdrawals were $25,000, the fund would grow. After the first withdrawal, the amount remaining would be $475,000. Since the fund earns 6% a year, just prior to the second withdrawal of $25,000 the amount in the fund would be $475,000(1.06) = $503,500, which is more than the $500,000 in the retirement account before the withdrawals started.

The young Pack members realized they were learning skills that would be beneficial throughout their lives.

HERKIMER'S FASCINATING FINANCIAL FACTS:

The Diner's Club credit card introduced in 1950 was really a charge card. A user did not have to pay any interest, but the entire balance was due at the end of each month. Diner's club made its money with an annual fee to cardholders and a surcharge to the merchant where the charge was made. Eight years later American Express developed a similar card. No interest was charged and the entire balance was due in full at the end of each month.

ACTIVITY SET FOR SESSION #26

1. **ALCAL** activity: The request references the spreadsheet produced by Glen and Roger. Use your calculator to check the amount remaining in the find at the start of the second year. The initial fund amount is $500,000, the amount withdrawn each year is $35,000, and the fund earns 6% a year:

To be evaluated	(500000-35000)(1.06)
Calculator entry	(500000-35000)*1.06
Number represented	

2. REFLECTION & COMMENT activity:

It used to be that people retired at 65 and died by 72. Seven years of retirement. Not anymore. Thanks to advances in medicine, quite a few folks live to a ripe old 85 or 90. Instead of seven years of retirement, many Americans can now look forward to 30 years of life after work. Thirty years! These words can strike terror into the heart ... or joy. Joy for those who've planned fun things to do with the last third of their time on earth and have saved enough money to pay for them. Terror for those who haven't planned or haven't saved. As it happens, far too many people would feel terror if they only knew how ill-prepared they are for 30 years of retirement. Or for 20 years. Or for 10.
(John Pierson, "Thirty Years of Retirement," *Fortune* (1993))

In the situations described in activities 3-7 fill in the tables to display fund amounts after the first three payments. (* Respond to this portion of the question if you can utilize spreadsheet skills or calculator skills to examine the fund for many years.)

3. Fund initial amount = $900,000
   Fund annual rate = 5.5%
   Amount withdrawn annually = $50,000

Year	Amount in fund	Amount withdrawn	Fund amount remaining	Fund amount with interest accumulation
1	$900,000			
2				
3				

   *Fund will last for _____ years.

4. Fund initial amount = $900,000
   Fund annual rate = 5.5%
   Amount withdrawn annually = $70,000

Year	Amount in fund	Amount withdrawn	Fund amount remaining	Fund amount with interest accumulation
1	$900,000			
2				
3				

   *Fund will last for _____ years.

5. Fund initial amount = $1,200,000
   Fund annual rate = 6.2%
   Amount withdrawn annually = $100,000

Year	Amount in fund	Amount withdrawn	Fund amount remaining	Fund amount with interest accumulation
1	$1,200,000			
2				
3				

   *Fund will last for _____ years.

6. Fund initial amount = $1,200,000
   Fund annual rate = 5.2%
   Amount withdrawn annually = $100,000

Year	Amount in fund	Amount withdrawn	Fund amount remaining	Fund amount with interest accumulation
1	$1,200,000			
2				
3				

*Fund will last for _____ years.

7. Fund initial amount = $800,000
   Fund annual rate = 5.4%
   Amount withdrawn annually = $30,000

Year	Amount in fund	Amount withdrawn	Fund amount remaining	Fund amount with interest accumulation
1	$800,000			
2				
3				

*Fund will last for _____ years.

$$$
$$

Session #27
A BIT MORE ON RETIREMENT PLANS

> *Anyone who lives within their means suffers from a lack of imagination.*
> -Oscar Wilde

Herkimer reminded the Pack that payments in the financial world are often done on a monthly basis. A retiree might want to have 12 monthly payments from a retirement fund rather than one annual payment. The fund interest rate might well be something other than a true annual rate.

"Suppose that a retirement fund contained $400,000 and it is desired to have monthly payments of $3,000," began Herkimer. "And, let's assume that the fund earns an annual rate of 6%, compounded monthly. Use your calculators and some good old pencil and paper and create a table displaying the first three months of payments. Let's see which team can get this on the chalkboard first."

It wasn't long before Frances and Darren took control of the chalkboard.

Fund Amt. = $400,000, $i^{(12)} = 6\%$, monthly withdrawals = $3,000. THE TRUE MONTHLY RATE IS $\frac{.06}{12} = .005$.

MONTH	AMT IN FUND	WITHDRAWN	REMAINING	ACCUM
1	$400,000	$3,000	$397,000	$397,000(1.005) = $398,985
2	$398,985	$3,000	$395,985	$395,985(1.005) = $397,965
3	$397,965	$3,000	$394,965	$394,965(1.005) = $396,940

"Good show," said Herkimer. "Now I actually have a spreadsheet that will produce a complete retirement payment schedule if you input the fund amount, an annual rate compounded monthly, and the monthly payments requested." He put this display on the overhead using the hidden row technique to get the schedule on one page:

152

	A	B	C	D	E	F
1	RETIREMENT FUND ANALYSIS….. Spreadsheet by HERKIMER					
2	User inputs FUND AMOUNT, INTEREST RATE, and AMOUNT WITHDRAWN EACH MONTH.					
3	Spreadsheet will provide length of time (up to 30 years) required for fund to be depleted.					
4						
5	$400,000	<--- INPUT INITIAL FUND AMOUNT				
6	6.00%	<--- INPUT ANNUAL RATE (COMPOUNDED MONTHLY) EARNED BY				
7	$3,000	<---INPUT AMOUNT YOU WISH TO WITHDRAW EACH MONTH.				
9						
10	Month #	Amount in Fund	Amount Withdraw	Amount remaining	Amount with accum. Interest	
11	1	$400,000	$3,000	$397,000	$398,985	
12	2	$398,985	$3,000	$395,985	$397,965	
13	3	$397,965	$3,000	$394,965	$396,940	
14	4	$396,940	$3,000	$393,940	$395,909	
15	5	$395,909	$3,000	$392,909	$394,874	
16	6	$394,874	$3,000	$391,874	$393,833	
17	7	$393,833	$3,000	$390,833	$392,788	
18	8	$392,788	$3,000	$389,788	$391,736	
19	9	$391,736	$3,000	$388,736	$390,680	
20	10	$390,680	$3,000	$387,680	$389,619	
21	11	$389,619	$3,000	$386,619	$388,552	
22	12	$388,552	$3,000	$385,552	$387,479	END YEAR 1
23	13	$387,479	$3,000	$384,479	$386,402	
24	14	$386,402	$3,000	$383,402	$385,319	
25	15	$385,319	$3,000	$382,319	$384,230	
26	16	$384,230	$3,000	$381,230	$383,137	
27	17	$383,137	$3,000	$380,137	$382,037	
28	18	$382,037	$3,000	$379,037	$380,932	
29	19	$380,932	$3,000	$377,932	$379,822	
30	20	$379,822	$3,000	$376,822	$378,706	
31	21	$378,706	$3,000	$375,706	$377,585	
32	22	$377,585	$3,000	$374,585	$376,458	
33	23	$376,458	$3,000	$373,458	$375,325	
34	24	$375,325	$3,000	$372,325	$374,187	END YEAR 2
35	25	$374,187	$3,000	$371,187	$373,042	
36	26	$373,042	$3,000	$370,042	$371,893	
37	27	$371,893	$3,000	$368,893	$370,737	
38	28	$370,737	$3,000	$367,737	$369,576	
39	29	$369,576	$3,000	$366,576	$368,409	
40	30	$368,409	$3,000	$365,409	$367,236	
41	31	$367,236	$3,000	$364,236	$366,057	
42	32	$366,057	$3,000	$363,057	$364,872	
43	33	$364,872	$3,000	$361,872	$363,682	
44	34	$363,682	$3,000	$360,682	$362,485	
45	35	$362,485	$3,000	$359,485	$361,282	
46	36	$361,282	$3,000	$358,282	$360,074	END YEAR 3
47	37	$360,074	$3,000	$357,074	$358,859	
48	38	$358,859	$3,000	$355,859	$357,639	
49	39	$357,639	$3,000	$354,639	$356,412	
50	40	$356,412	$3,000	$353,412	$355,179	
51	41	$355,179	$3,000	$352,179	$353,940	
52	42	$353,940	$3,000	$350,940	$352,694	
53	43	$352,694	$3,000	$349,694	$351,443	
54	44	$351,443	$3,000	$348,443	$350,185	
55	45	$350,185	$3,000	$347,185	$348,921	
56	46	$348,921	$3,000	$345,921	$347,651	
57	47	$347,651	$3,000	$344,651	$346,374	
58	48	$346,374	$3,000	$343,374	$345,091	END YEAR 4
59	49	$345,091	$3,000	$342,091	$343,801	
60	50	$343,801	$3,000	$340,801	$342,505	
61	51	$342,505	$3,000	$339,505	$341,203	
62	52	$341,203	$3,000	$338,203	$339,894	
63	53	$339,894	$3,000	$336,894	$338,578	
64	54	$338,578	$3,000	$335,578	$337,256	
65	55	$337,256	$3,000	$334,256	$335,927	
66	56	$335,927	$3,000	$332,927	$334,592	
67	57	$334,592	$3,000	$331,592	$333,250	
68	58	$333,250	$3,000	$330,250	$331,901	
69	59	$331,901	$3,000	$328,901	$330,546	
70	60	$330,546	$3,000	$327,546	$329,183	END YEAR 5
224						Hidden rows
225	215	$12,754	$3,000	$9,754	$9,802	
226	216	$9,802	$3,000	$6,802	$6,837	END YEAR 18
227	217	$6,837	$3,000	$3,837	$3,856	
228	218	$3,856	$3,000	$856	$860	
229	219	$860	$3,000	-$2,140	$0	

Retirement plan projections are only estimates and Herkimer wanted to make sure the Pack understood this. Interest rates can change over a period of years and amount withdrawals can vary. Referencing the previously monthly withdrawal schedule, he stressed that monthly withdrawals of $3,000 presents a different scenario than a single yearly withdrawal of $36,000. He got Glen and Roger to use their spreadsheet (Session #26) to illustrate a $400,000 fund with a true annual rate of 6% and an annual withdrawal of $36,000.

	A	B	C	D	E
1	Spreadsheet by GLEN and ROGER.				
2					
3	$400,000	<---Initial fund amount			
4	6%	<--True annual interest rate earned by fund			
5	$36,000	<---Amount withdrawn each year			
6				Fund amount	Fund amount with
7	Year	Amount in fund	Amount withdrawn	remaining	accumulated interest
8	1	$400,000	$36,000	$364,000	$385,840
9	2	$385,840	$36,000	$349,840	$370,830
10	3	$370,830	$36,000	$334,830	$354,920
11	4	$354,920	$36,000	$318,920	$338,055
12	5	$338,055	$36,000	$302,055	$320,179
13	6	$320,179	$36,000	$284,179	$301,229
14	7	$301,229	$36,000	$265,229	$281,143
15	8	$281,143	$36,000	$245,143	$259,852
16	9	$259,852	$36,000	$223,852	$237,283
17	10	$237,283	$36,000	$201,283	$213,360
18	11	$213,360	$36,000	$177,360	$188,002
19	12	$188,002	$36,000	$152,002	$161,122
20	13	$161,122	$36,000	$125,122	$132,629
21	14	$132,629	$36,000	$96,629	$102,427
22	15	$102,427	$36,000	$66,427	$70,412
23	16	$70,412	$36,000	$34,412	$36,477
24	17	$36,477	$36,000	$477	$506
25	18	$506	$36,000	($35,494)	($37,624)

"Keep in mind," began Herkimer, "that this spreadsheet assumes that the withdrawal of $36,000 comes at the beginning of the year. With monthly payments of $3,000, there is more left in the fund since the entire yearly amount of $36,000 is not immediately withdrawn in one payment. And, an annual rate of 6% compounded monthly is better than a true annual rate of 6%."

Pack members thought it would be interesting to create a table to comparing the two scenarios. Here's what they produced:

154

Amount in Fund	Initial Fund = $400,000. Annual Withdrawals = $36,000. True Annual Rate = 6%.	Initial Fund = $400,000. Monthly Withdrawals = $3000. Annual Rate Compounded Monthly = 6%.
After 1 year	$385,840	$387,479
After 2 years	$370,830	$374,187
After 3 years	$354,920	$360,074
After 4 years	$338,055	$345,091
After 5 years	$320,179	$329,183
Last complete withdrawal	Year #17	Month #218 (18 years, 2 months)

"Bear in mind," concluded Herkimer, "that many businesses and organizations have pension plans or retirement plan options for employees. If these are well managed, good folks can be reasonably well off in retirement. But let's not forget the power of compounding that we have discovered in our financial studies. As an individual, you can save money towards retirement. Small deposits can accumulate to large amounts over many years."

The Pack got the message.

HERKIMER'S FASCINATING FINANCIAL FACTS:

In 1959 Visa, known as Bank Americard, came out with the first card that allowed cardholders to create charges at multiple stores. This card, and others that followed, allowed customers to pay off the balance or carry it over many months at a monthly interest rate. Master Charge (later known as MasterCard) followed, as did many other banks as they realized the potential profits from credit cards. Basically, issuers of credit cards offered short term loans at high interest rates.

ACTIVITY SET FOR SESSION #27

1. **ALCAL** activity: The request references the spreadsheet produced by Herkimer. Use your calculator to check the amount remaining in the find at the start of the second month. The initial fund amount is $400,000, the amount withdrawn each month is $3,000, and the fund earns $i^{(12)} = 6\%$ a year:

To be evaluated	(400000-3000)(1+.06/12)
Calculator entry	(400000-3000)*(1+.06/12)
Number represented	

155

2. REFLECTION & COMMENT activity:

The average American will retire with just $57,000 at age 65 - that's after making more than $1.6 million over their lifetime. As a nation we've been saving less than 3%. In fact, in recent years, the person saving rate has hovered around a negative .02 percent! In other words, Americans are spending more than they're making.

(Forbes, *Other Comments* page, May 17, 1999)

3. Research retirement plans offered by various organizations. See what they involve. For instance, does the organization have a matching plan wherein it will match whatever amount an employee contributes to the plan each month?

4. What is an IRA? (The letters stand for Individual Retirement Account.) Research the tax benefits for those who establish IRA's. How does a Roth IRA differ from a traditional IRA?

5. In some retirement plans an employee has 5% of his salary put into a retirement plan and his company matches this amount. For instance, if an employee earns $3,000 a month, then (0.05)($3,000) = $150 is automatically deposited into a retirement account each month along with a matching $150 from the employer. A total of $300 is invested in some type of interest-earning account each month. Salaries change each year, and hence the amount invested each year varies over the course of many years. For future projection purposes, assume that an individual has saved $300 a month for 40 years. What would be the approximate amount in his retirement account at the end of the 40 years if the payments earned an annual rate of

(a) $i^{(12)} = 4\%$?   (b) $i^{(12)} = 5\%$?   (c) $i^{(12)} = 6\%$?   (d) $i^{(12)} = 7\%$?

6. Referencing the situation described in activity #3, the employee would have invested $150(12)(40) = $72,000 of his own money into the retirement fund. For each rate provided in activity #5, calculate how much of the 40-year retirement fund amount came from sources (matching payments by employer and interest) other than the employee.

7. Consider the four retirement fund amounts obtained in activity #5 and that money in the funds will continue to earn at the stated rate and that it is desired to withdraw $3,000 a month in retirement. The funds will last 150 months, 162 months, 178 months, and 200 months, respectively. That is, a full $3,000 monthly payment can be made over the indicated time period. (See if you can check this out with a spreadsheet or a calculator.) Keeping in mind that the individual invested only $72,000 of his own money (activity #6), find the total amount of the full payments of $3,000 that can be withdrawn in retirement.

$$
$$$

Session #28
SOME INSIGHT INTO BONDS

> *That some should be rich show that others may become rich, and hence is just encouragement to industry and enterprise.*
> -Abraham Lincoln

"When it comes to investing what do you read about most in the news?" Herkimer asked the Pack. Not surprisingly, the students mentioned the stock market. Stocks are exciting and newsworthy. The ups and downs of the stock markets, particularly the New York Stock Exchange, are scrutinized in daily newspapers and in television newscasts. Pack members were aware that one could make large amounts of money in the stock market; conversely, one could lose considerable money by making poor investment decisions. Stock prices fluctuate daily. An investment in a stock is not a fixed-income investment since one cannot determine exactly what it will be worth at any time in the future.

"Bonds are less exciting," said Herkimer, "but they do represent a fixed-income investment since you know exactly what you will get in the future. You can utilize what you have learned to do financial calculations relating to bonds. So that's what we are going to do."

He went on to explain that a bond is really a loan. When you purchase a bond, you are simply lending an amount (the purchase price) to a lender who agrees to pay you back with a future payment or payments. He placed this diagram on the overhead:

											$10,000
.	$800	$800	$800	$800	$800	$800	$800	$800	$800	$800	
Year 0	1	2	3	4	5	6	7	8	9	10	

"OK," said Herkimer, "this is a picture of a 10-year $10,000 bond with 8% annual coupons. If you buy this bond now, you are purchasing, actually lending, the seller money that will be paid back with the payments shown. The $10,000 is the *maturity value* of the bond and the $800 *coupons* simply represent annual interest payments that are each 8% of the maturity value. This is a fixed-income investment. What you see is what you get. The only question is what you are willing to pay for it."

From their previous experience with Herkimer, the Pack seemed to realize that the bond purchase price would simply be the present value of the displayed payments using the annual yield desired. This would mean finding the present value of a single payment of $10,000 and the present value of the series of $800 payments. Herkimer acknowledged that was indeed correct and challenged the Pack to calculate the purchase price for annual yields 6%, 8%, and 10%. It wasn't long before Janice and Carolyn began writing on the chalkboard.

10 YEAR \$10,000 BOND WITH 8% ANNUAL COUPONS:

TO YIELD 6%,

$$P = \$800\left[\frac{1-(1.06)^{-10}}{.06}\right] + \$10,000(1.06)^{-10} = \$11,472.$$

TO YIELD 8%,

$$P = \$800\left[\frac{1-(1.08)^{-10}}{.08}\right] + \$10,000(1.08)^{-10} = \$10,000.$$

TO YIELD 10%,

$$P = \$800\left[\frac{1-(1.10)^{-10}}{.10}\right] + \$10,000(1.10)^{-10} = \$8,771.$$

Herkimer drew everyone's attention to the second scenario on the board. "If your desired yield is the same as the coupon rate, you simply expect to pay the maturity value of the bond." He also pointed out that many bonds pay semi-annual coupons. In the above example, this would simply mean that two coupon payments of \$400 would be made each year rather than a single coupon payment of \$800 at the end of the year. "For simplicity, we'll stick with annual coupons," he said. "The annual yield would differ slightly if the coupons were semiannual, but not by very much."

"Do keep in mind," continued Herkimer, "that if you owned the bond described, you could sell it at any time. Remember, the payments are fixed. For instance, if I want to purchase it from you just after you have received the 6th coupon payment I would be buying 4 annual coupons of \$800 and a payment of \$10,000 in four years. If I wanted my purchase to yield 9.5%, what would I offer you for those payments?"

It didn't take the Pack long to come with the offer of \$9,519 obtained from the formula $\$800[1-(1.095)^{-4}]/.095 + \$10,000(1.095)^{-4}$.

This topic was ripe for spreadsheet activity. Herkimer challenged the Pack to find the purchase price of a \$100,000 twenty-five year bond with 7.5% annual coupons at yield rates ranging from 3% to 12%. He also wanted a graphics display related to the numerical output.

Brenda and Wayne met the challenge with this interactive spreadsheet:

158

	A	B	C	D	E	F	G	H
1	Spreadsheet by BRENDA and WAYNE							
2								
3	$100,000	INPUT bond maturity value			Maturity value =		$100,000	
4	25	INPUT number of years			Total coupon pymts =		$187,500	
5	7.50%	INPUT coupon rate			TOTAL =		$287,500	
6								
7	$7,500	Annual Coupon (calculated)						
8								
9	$287,500	Total bond payments						
10								
11			Total Gain (Total Payments					
12	Annual Yield	Purchase Price	- Purchase Price)					
13	3%	$178,359	$109,141					
14	4%	$154,677	$132,823					
15	5%	$135,235	$152,265					
16	6%	$119,175	$168,325					
17	7%	$105,827	$181,673					
18	8%	$94,663	$192,837					
19	9%	$85,266	$202,234					
20	10%	$77,307	$210,193		Chart below:			
21	11%	$70,524	$216,976		Bottom portion of each bar is purchase price.			
22	12%	$64,706	$222,794		Top portion of each bar is gain (Total			
23	13%	$59,685	$227,815		payments - purchase price).			
24								

"Beautiful," said Herkimer. "You've got the feel for bonds. Sometimes bonds represent good investments, sometimes not. It all depends on your investment objectives and the overall financial situation in the country."

ACTIVITY SET FOR SESSION #28

1. **ALCAL** activity: Use your calculator to check the chalkboard presentations made by Janice and Carolyn relating to a 10-year $10,000 bond with 8% annual coupons:

(a) The purchase price if the bond is purchased to yield 6%:

To be evaluated	$800(1-1.06^{-10})/.06+10000(1.06)^{-10}$
Calculator entry	800*(1-1.06^-10)/.06+10000*1.06^-10
Number represented	

(b) The purchase price if the bond is purchased to yield 10%:

To be evaluated	$800(1-1.10^{-10})/.10+10000(1.10)^{-10}$
Calculator entry	800*(1-1.10^-10)/.10+10000*1.10^-10
Number represented	

2. REFLECTION & COMMENT activity:

*The term **yield** is used to refer to the rate of return on a bond. This will vary from the coupon rate once a bond starts trading in the secondary market and ends up with a price that differs from the par value. If a bond has a fixed coupon interest rate (which is most commonly the case), the price of the bond has to adjust to keep the bond's yield in line with market conditions (particularly changes in interest rates) and changes in credit quality. Generally, interest rate risk has been the most important factor.*

(Philip Mattera, editor of *Public Bonds*)

3. Research the various types of bonds available for public purchase. These include *government bonds*, *municipal bonds*, and *corporate bonds*. There are three basic types of government bonds: *bills*, *notes*, and *bonds*. Distinguish between these types.

4. Stock prices rise when bond prices fall, and vice-versa. Research why this is so.

In activities 5-8, calculate the purchase price for each bond.

5. A $50,000 fifteen-year bond with 6.5% annual coupons purchased to yield 8%.

6. A $25,000 ten-year bond with 7% annual coupons purchased to yield 5.8%.

7. A $1,000,000 twenty-year bond with 6% annual coupons purchased to yield 7%.

8. A $75,000 thirty-year bond with 7.2% annual coupons purchased to yield 7.2%.

$$
$$

Session #29
YIELDS ON U.S. SAVINGS BONDS

> *Riches may enable us to confer favors, but to confer them with propriety and grace requires something that riches cannot give.*
> -Charles Caleb Colton

Some Pack members were familiar with U.S. Savings Bonds as a result of receiving such a bond as a gift when they were young children. Herkimer explained that such bonds represent loans to the federal government that are repaid in full, with interest, at a specified future date. He told the group that he would briefly introduces Series E and Series EE bonds, indicating that they were examples of zero-coupon bonds. That is, they paid no annual coupons and the purchaser would just receive the maturity value at a future time.

Herkimer stated that Series E bonds were issued from 1941 to 1980. Some are still in existence today. If you purchased a $100 bond, you paid 75% of the face value (in this case, $75) and redeemed the bond for $100 at a later date. It was popular to purchase $25 savings bonds for children. These bonds cost $18.75 and were, after a specified time, redeemed for $25. In general, one invested $3 to get a future return of $4. The yield rate was determined by the amount of time between purchase and redemption.

"You guys can figure out yields on this type of bond," said Herkimer. "Suppose," he continued, "a $100 Series E bond can be redeemed in 6 years. What is the yield rate? After you do that one, calculate the yield rate if it can be redeemed in 7 years, then the yield rate for 8 years."

Pack members remembered previous learning sessions. It wasn't long before the team of Frances and Valarie put the following computations on the chalkboard:

6 YEARS:
$$\$75(1+i)^6 = \$100 \Rightarrow (1+i)^6 = \frac{4}{3} \Rightarrow i = \left(\frac{4}{3}\right)^{\frac{1}{6}} - 1 = 4.91\%.$$

7 YEARS:
$$\$75(1+i)^7 = \$100 \Rightarrow i = \left(\frac{4}{3}\right)^{\frac{1}{7}} - 1 = 4.20\%.$$

8 YEARS:
$$\$75(1+i)^8 = \$100 \Rightarrow i = \left(\frac{4}{3}\right)^{\frac{1}{8}} - 1 = 3.66\%.$$

Herkimer then noted that some Series E bonds might be redeemable after a time period like 6 years and 6 months, or 6.5 years. To find the annual yield, it appears that the equation

$$\$3(1 + i)^{6.5} = \$4$$

would have to be solved. "What about that?" he asked.

After a brief time interval, Darren noted that $6.5 = 65/10$. He remembered from second-year algebra that $x^{65/10} = \sqrt[10]{x^{65}}$. "Hey, that financial equation does make some sense," he said. Darren did some computations on his calculator and then wrote this on the chalkboard:

$$\$3(1 + i)^{6.5} = \$4$$
$$\Rightarrow (1 + i)^{6.5} = 4/3$$
$$\Rightarrow 1 + i = \left(\frac{4}{3}\right)^{1/6.5}$$
$$\Rightarrow i = \left(\frac{4}{3}\right)^{1/6.5} - 1 = 4.525\%.$$

"Very good," said Herkimer. "The power of algebra definitely comes into play when working with mathematics involved with finance. With Series E bonds, an investment of $3 always returned $4. The annual yield is determined by the time interval between purchase and redemption. Now let's see who can produce a spreadsheet that will compute the annual yield on Series E bonds with maturities ranging from 6 years to 10 years using 3-month time intervals.

Stephen and Brenda were first to produce the requested spreadsheet. They even did some quick Internet searches to find out the possible denominations of Series E bonds.

	A	B	C	D	E	F	G
1	Spreadsheet by STEPHEN and BRENDA						
2	SERIES E BONDS						
3							
4	**BASIC CONCEPT: For every $3 you invest you get $4 back at a later date.**						
5							
6	**If x = number of years to redemption, yield rate = $(4/3)^{1/x} - 1$.**						
7							
8				x =Redemption	Yield	Yield rate (%)	
9	Face Value	Purchase Price		time (years)	rate	Two decimals	
10	$25	$18.75		6.00	0.049115	4.91%	
11	$50	$37.50		6.25	0.047105	4.71%	
12	$75	$56.25		6.50	0.045253	4.53%	
13	$100	$75.00		6.75	0.043541	4.35%	
14	$200	$150.00		7.00	0.041954	4.20%	
15	$500	$375.00		7.25	0.040478	4.05%	
16	$1,000	$750.00		7.50	0.039103	3.91%	
17	$5,000	$3,750.00		7.75	0.037818	3.78%	
18	$10,000	$7,500.00		8.00	0.036615	3.66%	
19				8.25	0.035486	3.55%	
20				8.50	0.034424	3.44%	
21				8.75	0.033424	3.34%	
22				9.00	0.032481	3.25%	
23				9.25	0.031589	3.16%	
24				9.50	0.030745	3.07%	
25				9.75	0.029945	2.99%	
26				10.00	0.029186	2.92%	
27							

164

After praising the work of Stephen and Brenda, Herkimer noted that Series E bonds were replaced by Series EE bonds in 1980. These bonds could be purchased for 1/2 of the face value. For instance, a $100 bond would cost $50 at the time of purchase. Again, the yield would be determined by the time between purchase and redemption.

HERKIMER'S FASCINATING FINANCIAL FACTS:

Savings bonds were first issued in 1935. The basic purpose was to provide a secure and attractive investment for people with modest incomes and to provide an additional source of money for the U.S. Treasury. Series E bonds (first issued in 1941) played a major role in financing government expenses during World War II. In 2001 the U.S. Treasury issued Series EE bonds designated as Patriot Bonds. Funds raised from these bonds contribute to the overall effort to fight the war on global terrorism.

ACTIVITY SET FOR SESSION #29

1. **ALCAL** activity: Use your calculator to check the spreadsheet computations produced by Stephen and Brenda relating to yields on Series E Bonds:

(a) The yield if the redemption time is 6 years:

To be evaluated	$(4/3)^{1/6} - 1$
Calculator entry	(4/3)^(1/6) - 1
Number represented	

(b) The yield if the redemption time is 9.5 years:

To be evaluated	$(4/3)^{1/9.5} - 1$
Calculator entry	(4/3)^(1/9.5) - 1
Number represented	

2. REFLECTION & COMMENT activity:

Buying U.S. savings bonds is patriotic; when you lend money to Uncle Sam, you're halping to finance the country's borrowing needs. It's good for America, and it's good for Americans, too. U.S. savings bonds are very safe: They are backed by the full faith of the U.S. government. When you spend $500 to buy a bond, it will never be worth less than $500.

(Dorothy Rose, Buying *U.S. Savings Bonds*, Bankrate.com)

3. Use the Internet to research the topic UNITED STATES SAVINGS BONDS.

4. Stocks and bonds are popular types of investments in our capitalistic society. How do these two investment types differ? Why is it frequently recommended that you have both types of investment in your overall investment portfolio?

5. A bond costing A dollars can be redeemed in N years for B dollars. In terms of A, B, and N, what is the annual yield for this bond?

6. As Herkimer stated in the session, Series EE bonds return $2 for each $1 invested.

   (a) If a Series EE bond can be redeemed after x years, what is the algebraic formula for the investment yield?

   (b) Calculate the yield on a Series EE bond for the indicated times from purchase to redemption:

# Years	16	16.5	17	17.75	18	19	20	21	22
Yield									

$$\$$$
$$\$$$

Session #30
YIELDS ON COUPON BONDS

A considerable amount of time had passed since the Pack had used the trial and error method to find solutions to financial equations, but the need to do this type of thing was about to present itself. Herkimer noted that the group had just used algebraic properties to derive yield rates for U.S. Saving bonds. However, coupon bonds presented a bigger challenge.

"OK," said Herkimer, "suppose a twenty-five year $100,000 bond with 6% annual coupons is offered to you for $92,000. If you purchase this bond, what is the yield rate on your investment? I'd like you to show me the equation that has to be solved to answer this question. And, then tell me how you would go about getting a solution."

After some discussion time, Valarie wrote this on the chalkboard while her teammate Stephen did some computations on his calculator:

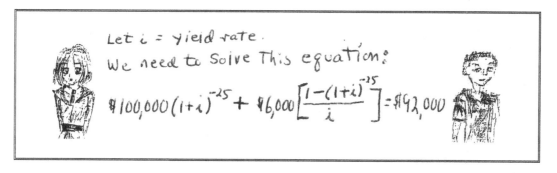

Let i = yield rate.
We need to solve this equation:

$$\$100,000(1+i)^{-25} + \$6,000\left[\frac{1-(1+i)^{-25}}{i}\right] = \$92,000$$

Stephen noted that it was just a case of "trapping a solution" as they had done in earlier sessions. He knew that the purchase price would be $100,000 if the yield was 6%. He had tried a 7% yield and found the process to be $100,000(1.07)^{-25} + $6,000[1-(1.07)^{-25}]/.07 = $88,346. Since this value was less than $92,000 the rate was trapped between 6% and 7%. Additional calculator work produced a price of $92,748 for 6.6% and $91,617 for 6.7%. Stephen concluded that the yield would be approximately 6.7%.

In the meantime the team of Janice and Wayne had been working on a powerful interactive spreadsheet that would more easily allow trapping of the yield with successive inputs. They demonstrated that the yield was indeed between 6.6% and 6.7%. They then used their interactive sheet to compute prices for rates between these two percentages and concluded that 6.67% was accurate to two decimal places. They also included a neat graphics display plotting yields and corresponding bond prices:

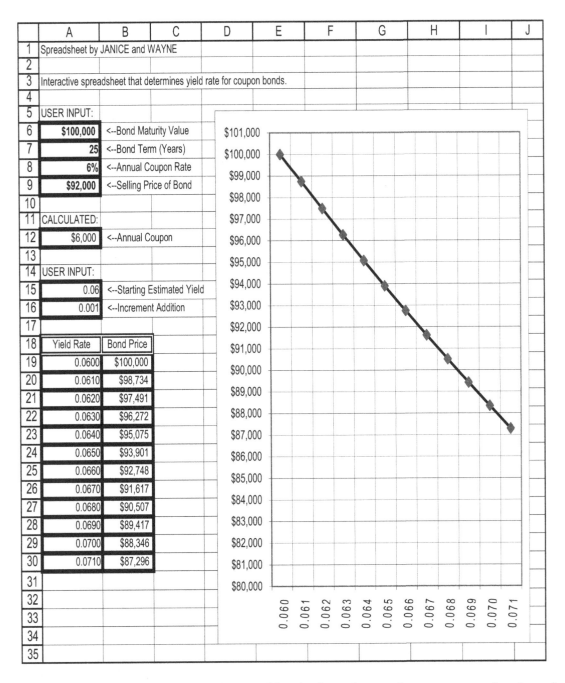

	A	B	C	D	E	F	G	H	I	J
1	Spreadsheet by JANICE and WAYNE									
2										
3	Interactive spreadsheet that determines yield rate for coupon bonds.									
4										
5	USER INPUT:									
6	$100,000	<--Bond Maturity Value								
7	25	<--Bond Term (Years)								
8	6%	<--Annual Coupon Rate								
9	$92,000	<--Selling Price of Bond								
10										
11	CALCULATED:									
12	$6,000	<--Annual Coupon								
13										
14	USER INPUT:									
15	0.06	<--Starting Estimated Yield								
16	0.001	<--Increment Addition								
17										
18	Yield Rate	Bond Price								
19	0.0600	$100,000								
20	0.0610	$98,734								
21	0.0620	$97,491								
22	0.0630	$96,272								
23	0.0640	$95,075								
24	0.0650	$93,901								
25	0.0660	$92,748								
26	0.0670	$91,617								
27	0.0680	$90,507								
28	0.0690	$89,417								
29	0.0700	$88,346								
30	0.0710	$87,296								
31										
32										
33										
34										
35										

"Really neat," responded Herkimer. "Simply from the graph you can see that the price function intersects the horizontal line y = $92,000 at approximately x = 0.0665, or about 6.7%." He went on to indicate that many modern graphics calculators have equation-solving features that would allow for a solution to many decimal places. One would simply have to enter the two functions:

$$y = 92000 \text{ and } y = 100000*(1+x)^{-25} + 6000*(1-(1+x)^{-25})/x$$

168

Setting the appropriate window, one could then note where the two graphs intersect. The calculator would provide the coordinates for the point of intersection.

HERKIMER'S FASCINATING FINANCIAL FACTS:

While most U.S. coins portray past Presidents, special coins to commemorate American people, places, events, and institutions are sometimes authorized by special acts of Congress and manufactured in limited quantities. These coins rarely circulate as a regular coin since collectors keep them reasoning that they will increase in value as time passes.

ACTIVITY SET FOR SESSION #30

1. **ALCAL** activity: Referencing the spreadsheet produced by Janice and Wayne, use your calculator to check the purchase price for a $100,000 twenty-five year bond with 6% annual coupons if

(a) the bond was purchased to yield 6.2%:

To be evaluated	$6000(1-1.062^{-25})/.062+100000(1.062)^{-25}$
Calculator entry	6000*(1-1.062^-25)/.062+100000*1.062^-25
Number represented	

(b) the bond was purchased to yield 7.1%:

To be evaluated	$6000(1-1.071^{-25})/.071+100000(1.071)^{-25}$
Calculator entry	6000*(1-1.071^-25)/.071+100000*1.071^-25
Number represented	

2. REFLECTION & COMMENT activity:

When prevailing interest rates rise, newly issued bonds typically offer higher yields to keep pace. When that happens, existing bonds with lower coupon rates become less competitive. That's because investors are unlikely to buy an existing bond offering a lower coupon rate unless they can get it at a lower price. Thus higher interest rates mean lower prices for existing bonds. Conversely, when interest rates fall, an existing bond's coupon rate becomes more appealing to investors, driving the price up.

(personal.fidelity.com, 2000)

169

In activities 3-6, write the formula that has to be solved to find the requested annual yield. Then use the trapping procedure to find the yield accurate to 2 decimal places (ie. 5.76%).

3. A 10-year $500,000 bond with 4% annual coupons is offered for $512,000. What is the annual yield?

4. A 25-year $1,000,000 bond with 5% annual coupons has a price of $970,000. Calculate the annual yield.

5. An $80,000 five-year bond with 6% annual coupons has a listed price of $76,000. If you purchase this bond, what is your annual yield?

6. Calculate the annual yield on a $250,000 twenty-year bond with 7% annual coupons that has a listed $260,000 purchase price.

----

7. A 25-year $200,000 bond with 6.5% annual coupons that was issued 20 years ago is being offered at $180,000. If you purchase this bond, what is your annual yield? [You are purchasing 5 of the coupons along with the maturity value of $200,000.]

$$$
$$

HERKIMER'S QUICK QUIZ
SESSIONS 26-30.

Find the correct response for
each question in the 25 cells
at the bottom of the page.

1. What is the purchase price of a twenty-year $100,000 bond with 6% annual coupons purchased to yield 8%?

2. A 25-year $1,000,000 bond has 6.5% annual coupons. If one purchases this bond, what is the total income that would be received over the 25-year bond term?

3. A Series E Savings Bond that could be redeemed in five years and nine months had an annual yield of _____.

4. What equal annual deposits earning a 6.3% annual yield would produce a retirement fund of $600,000 in 35 years?

5. Each year an annual withdrawal of $45,000 is made from a $600,000 retirement fund. If the fund earns an annual interest rate of 5.9%, how much does it contain just before the second withdrawal?

6. Series EE Savings Bonds returned $2 for a purchase price of $1. What is the annual yield if the redemption time is 15 years and 9 months?

7. A $20,000 ten-year bond with 10% annual coupons purchased to yield 10% has a purchase price of _____

8. A zero-coupon bond is redeemable for Y dollars in T years. If this bond sells for P dollars, what is the annual yield?

9. Each year the interest from an $800,000 retirement fund is withdrawn by a retiree. If the fund earns an annual interest rate of 6%, how much does the fund contain just after the 10th annual withdrawal?

10. What is the annual yield on a 20-year $500,000 bond with annual coupons of 6% purchased for $472,000?

	A	B	C	D	E
1	$587,745	4.88%	$(P/Y)^{T} - 1$	6.51%	$(Y/T)^{1/P} - 1$
2	$1,500,000	$2,625,000	$22,755	$2,287,345	6.02%
3	8.27%	$(Y/P)^{1/T} - 1$	$602,387	$522,543	$80,364
4	$73,800	$29,544	$17,439	4.50%	$800,000
5	$4,751	$1,995,000	$20,000	$546,934	5.13%

Session #31
AVOID PONZI SCHEMES

> *My problem lies in reconciling my gross habits with my net income.*
> -Errol Flynn

"What is a Ponzi scheme?" asked Herkimer. While some Pack members had heard the phrase before, none knew what a Ponzi scheme really was. Curiosity was present within the group, so Herkimer decided to expand on the topic.

He explained that the promoter of a Ponzi scheme takes in investors' money for shares in a fund and within a short amount of time pays these investors impressive "profits." However these so-called profits, unbeknownst to the investors, are simply a portion of the investors' own money. These impressive profits attract other investors who buy shares in the fund. People with money in the fund recommend it to friends who are impressed with this investment paying unusually high returns. The perpetrator may succeed in taking in large sums of money for a period of time unless someone gets suspicious of the source of the profits. However a Ponzi scheme is doomed for eventual failure since the promoter cannot generate an endless supply of new investors to generate cash for profit payments.

Herkimer continued on to explain that in a Ponzi scheme there will come a time when the payments of profits will cease. The perpetrator, if still around, will offer a multitude of excuses as to why profits are not being paid.

"What happens if an investor wants to sell his shares to get back the money he invested?" asked Carolyn.

"Such an investor will hear more excuses about why that can't be done," replied Herkimer. "Once profits cease and you can't get a return of principal invested, you can be pretty sure you are caught up in something that's not so good."

Herkimer was about to place on the overhead a spreadsheet illustrating a simple Ponzi scheme. "This is greatly overly simplified," he said, "but it will illustrate why a Ponzi scheme is doomed to fail." He went on to explain that in this simple example investors could buy shares in this fund for $100 on the first of each month. On the last day of the month each shareholder received a profit of $10 per share.

"Wow, a 10% per month return would sure seem impressive," commented Glen.

"You bet," replied Herkimer. "But remember that investors are simply getting back some of their own money. Note carefully what happens when the number of new shares decreases as the number of new investors declines."

Herkimer placed the spreadsheet on the overhead projector:

	A	B	C	D	E	F
1	**PONZI SCHEME DEMONSTRATION**					
2						
3	Shares are purchased on the first day of each month.					
4	Profits are paid on the last dayof each month.					
5						
6	**$100**	<--- Cost of share				
7	**10%**	<---Monthly profit rate				
8	**$10**	<----Monthly profit per share				
9						
10	Date	New Shares	Shares Outstanding	Added to Fund	Profits paid	Amount in Fund
11	**1-Jan**	60	60	**$6,000**		**$6,000**
12	31-Jan				*$600*	$5,400
13	**1-Feb**	90	150	**$9,000**		**$14,400**
14	28-Feb				*$1,500*	$12,900
15	**1-Mar**	100	250	**$10,000**		**$22,900**
16	31-Mar				*$2,500*	$20,400
17	**1-Apr**	110	360	**$11,000**		**$31,400**
18	30-Apr				*$3,600*	$27,800
19	**1-May**	120	480	**$12,000**		**$39,800**
20	31-May				*$4,800*	$35,000
21	**1-Jun**	10	490	**$1,000**		**$36,000**
22	30-Jun				*$4,900*	$31,100
23	**1-Jul**	4	494	**$400**		**$31,500**
24	31-Jul				*$4,940*	$26,560
25	**1-Aug**	3	497	**$300**		**$26,860**
26	31-Aug				*$4,970*	$21,890
27	**1-Sep**	2	499	**$200**		$21,890
28	30-Sep				*$4,990*	$16,900
29	**1-Oct**	1	500	**$100**		$16,900
30	31-Oct				*$5,000*	$11,900
31	**1-Nov**	0	500	**$0**		$11,900
32	30-Nov				*$5,000*	$6,900
33	**1-Dec**	0	500	**$0**		$6,900
34	31-Dec				*$5,000*	$1,900

The Pack noted that in this simple version of a Ponzi scheme the situation began to unravel after six months. If the perpetrator continued to pay the 10% a month profits, the fund would consist of $1,900 after one year during which a total 500($100) = $50,000 had been invested. "Can you imagine what would happen if many investors in this fund simply wanted to cash in their shares?" asked Herkimer.

Pack members were quick to notice that with only $1,900 in the fund at the end of December, it would be difficult to pay shareholders $5,000 in profits without many more investors during the upcoming January month. This Ponzi scheme is indeed doomed and many good, but gullible folks would lose lots of money.

"There are legitimate funds in which people can invest," continued Herkimer, "but you want to be sure you examine investment opportunities carefully. If you are being asked to purchase shares of a fund that has a very short history of paying high profits generated from unexplained or vaguely-explained sources you may well be being enticed into a Ponzi scheme. Research has indicated that a cleverly constructed Ponzi scheme can last about 18 months before investors get suspicious due to "failure to continue to pay profits" excuses, but some have been kept afloat for many years and have resulted in losses in the millions of dollars."

Herkimer concluded by dredging up the old "If it's too good to be true, it probably is," saying. The Pack had no trouble relating it to the Ponzi scheme concept.

HERKIMER'S FASCINATING FINANCIAL FACTS:

Paper currency is designed so that it will not tear easily. On average, a United States bill can undergo 4000 double folds (first forward, then backward) before it tears.

ACTIVITY SET FOR SESSION #31

1. **ALCAL** activity: Suppose a Ponzi scheme initially had 100 investors and had to increase its participation by 50% every month. Use your calculator to evaluate $100*1.5^X$ for various values of X to determine the number of months until
   - (a) 1,000,000 participants would be needed;
   - (b) 100,000,000 participants would be needed.

2. REFLECTION & COMMENT activity:

   The problem with a Ponzi investment scheme is that it is difficult to sustain this game very long because to continue paying the promised profits to early investors you need an ever-larger pool of later investors. The idea behind this type of swindle is that the con-man collects his money from his second our third round of investors and then beats it out of town before anyone else comes around to collect. These schemes typically only last weeks, or months at most. There is of course always the temptation to stay around just a little longer to collect another round of investments.

   (Larry DeWitt, *Social Security Online*, January 2009)

3. Research the history of Ponzi schemes. Among other things, how did the name Ponzi become associated with the illegal activity?

4. In the early years of the 21st Century a $50 billion dollar Ponzi scheme with thousands of clients was masterminded by Bernard Madoff. Research this specific Ponzi scheme. How did Madoff mange to fool many intelligent and well-known people? How did Madoff get so many people to invest money with him? List some of the famous people who lost considerable amounts of money. What eventually led investors to realize they were being scammed?

5. Research the concept of a ***pyramid scheme***. These differ from Ponzi schemes and rely primarily on participants having an ignorance of basic mathematics. Participants are promised large rewards for putting up a certain amount of money and then recruiting the next level of members. In 2009 the world's population was approximately 6,800,000,000 (6.8 billion). The person at the top of the pyramid scheme (Level 0) might recruit 12 people to "invest" money, and each of these 12 (Level 1) would recruit 12 people (Level 2), etc. Assuming that each investor contacts 12 people who have not previously been contacted, at what level would every person in the world been contacted? Continue the powers of 12 displayed below to obtain an answer. [In 2009 the world's population was approximately 6,800,000,000 (6.8 billion).]

$$12^0 = 1 \ \text{(Level 0)}$$

$$12^1 = 12 \ \text{(Level 1)}$$

$$12^2 = 144 \ \text{(Level 2)}$$

$$12^3 = 1,728 \ \text{(Level 3)}$$

6. Complete the following table that follows the pattern of the one demonstrated in the session #31. In this Ponzi scheme the cost of a share is $10,000. Shares can be purchased on the first of each month and profits of 12% per share are paid on the last day of each month. In this scenario, the fund starts with 20 shares and the number of new shares doubles each month for the first 6 months. Thereafter, no new shares are purchased.

$10,000	<--- Cost of share
12%	<---Monthly profit rate
$1,200	<----Monthly profit per share

Date	New Shares	Shares Outstanding	Added to Fund	Profits paid	Amount in Fund
1-Jan	20	20	$200,000		$200,000
31-Jan				$24,000	$176,000
1-Feb	40	60	$400,000		$576,000
28-Feb				$72,000	$504,000
1-Mar	80				
31-Mar					
1-Apr	160				
30-Apr					
1-May	320				
31-May					
1-Jun	640				
30-Jun					
1-Jul	0				
31-Jul					
1-Aug	0				
31-Aug					
1-Sep	0				
30-Sep					
1-Oct	0				
31-Oct					
1-Nov	0				
30-Nov					
1-Dec	0				
31-Dec					

$$$
$$$

176

Session #32
PACK MEMBERS DO INDIVIDUAL PROJECTS

He who loses money, loses much; he who loses a friend, loses much more; he who loses faith, loses all.
-Eleanor Roosevelt

Pack members knew that the learning sessions with Herkimer were almost over. As he had done with the group during the study of statistics, Herkimer decided to have the students do some individual research on ten specific financial topics. By a random selection process, each student would be assigned a topic. After providing time to do whatever research was needed along with appropriate charts and tables, each Pack member would be provided with ample time and resources to make his or her presentation to the group.

Looking over the list of topics the Pack realized that, for the most part, they simply represented extensions of financial concepts they had learned under Herkimer's guidance. They looked forward to the challenge being presented to them.

The topics were identified and labeled 1 through 10, as indicate on the following table. Herkimer put the names of the 10 students in a bag, mixed them thoroughly, and then drew names, one by one, to assign a student to a specific topic.

The assignment outcome is presented in the following table:

Project #	Project	Randomly assigned to:
1	THE RULE OF 72: This rule allows one to produce a fairly accurate estimate of the number of years required for money to double at a specified interest rate.	BRENDA
2	PAYDAY LOANS: These loans are quite common. However, interest rates are astronomical and one would be probably be wise to avoid this type of money exchange.	VALARIE
3	START TO SAVE EARLY: People frequently wait to long to begin a savings plan. The sooner one can put money into a good savings plan, the better. If one starts early, small amounts invested early earn more than larger amounts invested later.	WAYNE
4	CONTINUOUS INTEREST: Interest is sometimes compounded continuously. This relatively simple process involves the amazing number e, the base of natural logarithms.	DARREN
5	ADVANTAGE OF EXTRA MORTGAGE PAYMENTS: A few extra dollars each month can save thousands of dollars in total payments over the life of a mortgage. Since interest is calculated on the outstanding principal after each payment, extra dollars can reduce the principal a borrower owes.	FRANCES

177

6	REFINANCING: If interest rates drop, the possibility of refinancing a mortgage or a general loan might be worthy of consideration for a variety of reasons. However, one should clearly understand all financial obligations involved before attempting to refinance any debt.	CAROLYN
7	BOND AMORTIZATION SCHEDULES: The holder of a bond has an asset that increases or decreases in value over a period of time. For a variety of reasons, the bond holder needs to know the value of his/her investment after annual interest (coupons) have been paid.	ROGER
8	INFLATION CONCERNS IN RETIREMENT: Good retirement planning involves considering the effects of inflation on money that is being saved for use in future years. One must realize that $1 will probably buy more today that it will many years from now.	STEPHEN
9	TEASER RATES: A teaser rate is a very low interest rate that is offered to encourage one to sign up and use a particular credit card. The problem is that the rate is only for a limited amount of time, often six months or a year. After that period the situation can change drastically.	GLEN
10	ADJUSTABLE RATE MORTGAGES (ARMs): If one understands the terms and situations related to a specific mortgage with an adjustable interest rate, such a loan might be worth considering. Unfortunately, many people don't see beyond the initial low monthly payment.	JANICE

HERKIMER'S FASCINATING FINANCIAL FACTS:

On the first day of a 31-day month you put a penny in a basket. Each day thereafter you double the number of pennies you put in the basket. That is, the number of pennies you deposit follow the sequence 1, 2, 4, 8, 16, …. for 31 days. If you actually did this the basket would contain 2,147,483,647 pennies at the end of the month. You would have $21,474,836.47. (Needless to say, you would probably need more than one basket to hold the pennies.)

ACTIVITY SET FOR SESSION #32

1. **ALCAL** activity:

(a) Consider the number 43^2. Is the number $4^9 = 262,144$ or is it
$64^2 = 4096$. Computer 4^3^2 on your calculator. In this entry interpreted as
(4^3)^2 or 4^(3^2)? (NOTE: Not all calculators are consistent in representing a
number like this.)

(b) The largest number that can be represented with exactly three digits combined with
number operations is 9^(9^9). What happens when you attempt to compute this
number on your calculator?

2. REFLECTION & COMMENT activity:

*According to the JumpStart Coalition for Financial Literacy, the average high school
student lacks basic financial know-how. Balancing a checkbook, investing, and credit card
do's and don'ts are skills many teenagers neglect to learn, simply ignore or were never
taught in school or at home. Upon graduation, these otherwise smart kids get themselves
into financial trouble before they realize they aren't living within their means.*

(*Good Neighbor,* State Farm Magazine, Summer 2009*)*

$$\$$$
$$\$$$

Session #33
BRENDA'S PROJECT: THE RULE OF 72

> *The chief value of money lies in the fact that one lives in a world in which it is overestimated.*
> -H. L. Mencken

Brenda explained that her research indicated that the *Rule of 72* could be used to estimate how long it would take a single investment to double in value at various interest rates. "It's simple," she said. "You multiply the interest rate by 100 and then divide what you get into 72. This provides a good approximation for the number of years required for doubling."

Brenda had prepared a transparency to demonstrate the *Rule of 72*.

$$i = 5\% : \text{Estimated doubling Time} = \frac{72}{100(.05)} = \frac{72}{5} = 14.4 \ (\text{years}).$$
$$\text{Actual Time}: (1.05)^x = 2 \Rightarrow x = \frac{\log 2}{\log 1.05} = 14.21 \ (\text{years}).$$

$$i = 10\% : \text{Estimated doubling Time} = \frac{72}{100(.10)} = \frac{72}{10} = 7.2 \ (\text{years}).$$
$$\text{Actual time}: (1.10)^x = 2 \Rightarrow x = \frac{\log 2}{\log (1.10)} = 7.27 \ (\text{years}).$$

$$i = 12\% : \text{Estimated doubling Time} = \frac{72}{100(.12)} = \frac{72}{12} = 6 \ (\text{year}).$$
$$\text{Actual time}: (1.12)^x = 2 \Rightarrow x = \frac{\log 2}{\log (1.12)} = 6.12 \ (\text{years}).$$

Brenda also prepared a spreadsheet comparing the estimated doubling time to the actual time. She indicated that in terms of error percentage, the *Rule of 72* is reasonably accurate up to about 20%. Beyond that point the error is greater than 5% and continues to increase. Herkimer asked the Pack to not display the row and column headings in the spreadsheets they prepared for their respective projects. Brenda and others adhered to that request.

Spreadsheet by BRENDA				
Topic: **The RULE OF 72.**				

The RULE OF 72 allows one to estimate fairly accurately how many years are required for money to double at a specified annual interest rate. Given an annual interest rate, multilpy it by 100 and divde the product obtained into 72. The number obtained is the approximate number of years required for money to double at the indicated rate.

A DEMONSTRATION (This portion of the spreadsheet is interactive):

INPUT annual interest rate ---------------------->			5.00%	
Multiply this rate by 100, obtaining-------------->			5	
Divide	5	into 72, producing	14.40	(# years for money to double)
Using the formula				
Log(2)/Log(1+x) with x =	5.00%	yields	14.21	(years)

Table comparing Log(2)/Log(1+x) with 72/(100x) where x is an annual interest rate.
% Error = |Actual - Estimate|/Actual.

x	Log(2)/Log(1+x)	72/(100x)	% Error
1.00%	69.66	72.00	3.36%
2.00%	35.00	36.00	2.85%
3.00%	23.45	24.00	2.35%
4.00%	17.67	18.00	1.85%
5.00%	14.21	14.40	1.36%
6.00%	11.90	12.00	0.88%
7.00%	10.24	10.29	0.40%
8.00%	9.01	9.00	0.07%
9.00%	8.04	8.00	0.54%
10.00%	7.27	7.20	1.00%
11.00%	6.64	6.55	1.45%
12.00%	6.12	6.00	1.90%
13.00%	5.67	5.54	2.34%
14.00%	5.29	5.14	2.78%
15.00%	4.96	4.80	3.22%
16.00%	4.67	4.50	3.64%
17.00%	4.41	4.24	4.07%
18.00%	4.19	4.00	4.49%
19.00%	3.98	3.79	4.90%
20.00%	3.80	3.60	5.31%

While researching the *Rule of 72*, Brenda was curious if there were other approximation formulas similar to it. She readily found some on the internet. The time for money to triple can be approximated by replacing 72 by 114; for money to increase 4-fold, replace 72 by 144; for money to increase 5-fold, replace 72 by 167. She displayed an overhead transparency reflecting these discoveries:

181

If x is a true annual interest rate, then

$(1+x)^n = 3 \Longrightarrow nLog(1+x) = Log(3) \Longrightarrow n = Log(3)/Log(1+x)$. This value can be approximated by $114/(100x)$.

$(1+x)^n = 4 \Longrightarrow nLog(1+x) = Log(4) \Longrightarrow n = Log(4)/Log(1+x)$. This value can be approximated by $144/(100x)$.

$(1+x)^n = 5 \Longrightarrow nLog(1+x) = Log(5) \Longrightarrow n = Log(5)/Log(1+x)$. This value can be approximated by $167/(100x)$.

Brenda concluded with this spreadsheet showing the relative accuracy of the approximations.

Second spreadsheet by BRENDA

Demonstration of estimation formulas related to money growth.

	MONEY TRIPLES			MONEY INCREASES 4-FOLD			MONEY INCREASES 5-FOLD		
	Actual time(Yrs)	Est. time		Actual time(Yrs)	Est. time		Actual time(Yrs)	Est. time	
x	Log(3)/Log(1+x)	114/(100x)	Error	Log(4)/Log(1+x)	144/(100x)	Error	Log(5)/Log(1+x)	167/(100x)	Error
1%	110.41	114.00	3.25%	139.32	144.00	3.36%	161.75	167.00	3.25%
2%	55.48	57.00	2.74%	70.01	72.00	2.85%	81.27	83.50	2.74%
3%	37.17	38.00	2.24%	46.90	48.00	2.35%	54.45	55.67	2.24%
4%	28.01	28.50	1.75%	35.35	36.00	1.85%	41.04	41.75	1.74%
5%	22.52	22.80	1.26%	28.41	28.80	1.36%	32.99	33.40	1.25%
6%	18.85	19.00	0.77%	23.79	24.00	0.88%	27.62	27.83	0.77%
7%	16.24	16.29	0.30%	20.49	20.57	0.40%	23.79	23.86	0.29%
8%	14.27	14.25	0.17%	18.01	18.00	0.07%	20.91	20.88	0.18%
9%	12.75	12.67	0.64%	16.09	16.00	0.54%	18.68	18.56	0.64%
10%	11.53	11.40	1.10%	14.55	14.40	1.00%	16.89	16.70	1.10%
11%	10.53	10.36	1.55%	13.28	13.09	1.45%	15.42	15.18	1.56%
12%	9.69	9.50	2.00%	12.23	12.00	1.90%	14.20	13.92	2.01%
13%	8.99	8.77	2.44%	11.34	11.08	2.34%	13.17	12.85	2.45%
14%	8.38	8.14	2.88%	10.58	10.29	2.78%	12.28	11.93	2.89%
15%	7.86	7.60	3.32%	9.92	9.60	3.22%	11.52	11.13	3.32%
16%	7.40	7.13	3.74%	9.34	9.00	3.64%	10.84	10.44	3.75%
17%	7.00	6.71	4.17%	8.83	8.47	4.07%	10.25	9.82	4.17%
18%	6.64	6.33	4.58%	8.38	8.00	4.49%	9.72	9.28	4.59%
19%	6.32	6.00	5.00%	7.97	7.58	4.90%	9.25	8.79	5.00%
20%	6.03	5.70	5.40%	7.60	7.20	5.31%	8.83	8.35	5.41%
21%	5.76	5.43	5.81%	7.27	6.86	5.71%	8.44	7.95	5.81%
22%	5.52	5.18	6.21%	6.97	6.55	6.11%	8.09	7.59	6.21%
23%	5.31	4.96	6.60%	6.70	6.26	6.51%	7.77	7.26	6.61%
24%	5.11	4.75	6.99%	6.44	6.00	6.90%	7.48	6.96	7.00%
25%	4.92	4.56	7.38%	6.21	5.76	7.28%	7.21	6.68	7.38%
26%	4.75	4.38	7.76%	6.00	5.54	7.67%	6.96	6.42	7.77%
27%	4.60	4.22	8.14%	5.80	5.33	8.05%	6.73	6.19	8.14%
28%	4.45	4.07	8.51%	5.62	5.14	8.42%	6.52	5.96	8.52%
29%	4.31	3.93	8.88%	5.44	4.97	8.79%	6.32	5.76	8.89%
30%	4.19	3.80	9.25%	5.28	4.80	9.16%	6.13	5.57	9.25%

182

The Pack concluded that Brenda had done a thorough job with her project.

HERKIMER'S FASCINATING FINANCIAL FACTS:

In 1997 the 50 States Quarters Program Act redesigned the back side of quarters to depict emblems for each of the 50 states. Each year from 1999 through 2008, coins commemorating five states were issued in the order of which each state joined the Union. These quarters are now in general circulation.

ACTIVITY SET FOR SESSION #33

1. **ALCAL** activity: Use your calculator to check the following computations on the first spreadsheet produced by Brenda for a 13% annual interest rate:

(a) The exact time (years) for money to double:

To be evaluated	Calculator entry	Number represented
log(2)/log(1.13)	log(2)/log(1.13)	

(b) The approximate time (years) for money to double:

To be evaluated	Calculator entry	Number represented
72/(100*.13)	72/(100*.13)	

2. REFLECTION & COMMENT activity:

In the case of refinancing, most people only care about the immediate, short-term savings, failing to see the real costs they will incur over the long run.

(Joshua Kennon, *Beginners Investment Guide*)

3. Use the Internet to research why the *Rule of 72* provides good approximations for doubling times for annual interest rates of 20% or less.

4. Find a solution for the equation $72/(100x) = \text{Log}(2)/\text{Log}(1+x)$ accurate to 4 decimal places. In other words, for what interest rate is the *Rule of 72* approximation extremely close to the actual interest rate for money doubling time? And what is the time required for money to double at this rate?

5. Fill in the following table to illustrate that the *Rule of 72* approximation becomes progressively less accurate for annual rates exceeding 20%:

x = annual interest rate	Actual Time Required (Yrs) Log(2)/Log(1+x)	Estimated time by *Rule of 72* 72/(100x)	% Error
0.25 (25%)			
0.30 (30%)			
0.35 (35%)			
0.40 (40%)			
0.50 (50%)			

6. Examine the second spreadsheet produced by Brenda. For what interest rate are the approximation formulas most accurate? Is it true that the approximation formulas become less accurate as the interest rate increases beyond 20%?

$$\$$$
$$\$$$

Session #34
VALARIE'S PROJECT: PAYDAY LOANS

> *If you want to know what God thinks of money, just look at the people He gave it to.*
> -Dorothy Parker

Pack members were somewhat familiar with the concept of a payday loan since they had seen ads for them on television and the internet, and heard ads for them on the radio. Valarie began by noting that she was extremely surprised by what she found during her research on payday loans. "Among other things," she said, "the interest rates on these loans are astronomical and laws relating to them vary from state to state. I concluded that you definitely want to make every effort to not get involved with these things."

Valarie explained that a payday loan is usually for a small amount over a short time period that is usually two weeks. Borrowers use them to cover expenses until the next payday. Typical loans usually range from $100 to $500. On average, these loans cost $15 for each $100 borrowed. A typical transaction has the borrower writing the lender a postdated check for $115 in exchange for $100 in cash. On payday, the borrower may return to the lender with $115 in cash and have the written check returned to him, or the lender processes the check in the traditional manner.

"This all sounds innocent enough," said Valarie, "but look at this overhead I have prepared." She placed a neatly typed transparency on the projector. Herkimer and the Pack noted financial symbolism from previous sessions.

Loan = $100.

Interest payment for 2-week period = $15.
Interest rate for 2-week period = 15%.

Number of 2-week periods in 1 year = 26.

$i^{(26)}/26 = 15\%$ ==> $i^{(26)} = (26)(15\%) = 390\%.$
The nominal annual rate is 390% compounded 26 times per year.

True annual rate = $(1 + i^{(26)}/26)^{26} - 1 = (1.15)^{26} - 1 = 3686\%.$

"Do you realize," Valarie continued, "that if your bank paid a true annual rate of 3600% then for each $1 you invested you would have $1(1 + 36) = $37 after one year?

Valarie then produced a spreadsheet displaying the true annual interest rates on a two-week $100 payday loan for loan fees ranging from $10 to $20.

Spreadsheet by VALARIE				
Topic: PAYDAY LOANS				
These are small amount loans for which borrowers are charged a loan fee.				
This table displays the true annual rate received by lenders.				
The second column is the **loan fee charged for each $100 borrowed**.				
Amount Borrowed	Loan Fee	True Rate for 2-Week Period	Nominal Annual Rate $i^{(26)}$	True Annual Rate
$100	$10	10%	260%	1092%
$100	$11	11%	286%	1408%
$100	$12	12%	312%	1804%
$100	$13	13%	338%	2299%
$100	$14	14%	364%	2917%
$100	$15	15%	390%	3686%
$100	$16	16%	416%	4641%
$100	$17	17%	442%	5827%
$100	$18	18%	468%	7295%
$100	$19	19%	494%	9109%
$100	$20	20%	520%	11348%

Pack members were startled by the annual yields related to payday loans.

Valarie then explained the *roll-over* feature of these loans. Her research indicated that rolling over a loan is not allowed in some states, but where allowed the process goes like this and does illustrate an important financial concept. Consider the $100 payday loan with a fee of $15. If the borrower cannot repay the $100 at the end of two weeks, he *can roll* over the loan for another two weeks by paying a second $15 loan fee. If the borrower continued rolling over for a year before repaying the loan, he would pay a total of $100 + 26($15) = $490 for the original $100 loan. "Now here's the catch," said Valarie. "If the borrower paid the entire $490 at the end of one year, the true annual rate would 390%. The rate would come from solving the equation $100(1 + i) = $490. This solution is i = 3.9, or 390%. But the borrower does not pay the $390 all at once at the end of the year, but by increments of $15 every two weeks. He doesn't have the full use of the $390 for the entire year. This translates to an annual yield of 3686%, as my spreadsheet table indicates."

"A wonderful presentation," responded Herkimer. It was generally agreed that payday loans are very expensive and should be avoided despite the variety of ads that promote them in the media.

ACTIVITY SET FOR SESSION #34

1. **ALCAL** activity: Use your calculator to check the true annual rates displayed in Valarie's spreadsheet:

   (a) Loan fee is $10 per 2-week period for each $100 borrowed. In this situation $i^{(26)} = 260\%$:

To be evaluated	Calculator entry	Number represented
$(1+2.60/26)^{26}-1$	(1+2.60/26)^26-1	

   (b) Loan fee is $20 per 2-week period for each $100 borrowed. In this situation $i^{(26)} = 520\%$:

To be evaluated	Calculator entry	Number represented
$(1+5.20/26)^{26}-1$	(1+5.20/26)^26-1	

2. REFLECTION & COMMENT activity:

   People who resort to payday lending are typically low-income people with few assets, as they are people who are least able to secure normal, low-interest-rate forms of credit. Since the payday lending operations charge such high interest rates and do nothing to encourage savings or asset accumulation, they have the effect of depleting the assets of low-income communities.

   (Howard Jacob Karger, The Social Policy Journal, 2004))

3. Research the concept *payday loans* on the internet. Identify some of the problems borrowers often encounter with these loans. How do states differ in their laws relating to these loans?

187

4. What do payday lenders require from borrowers before a loan will be approved. (In general payday lenders require considerably less that other types of lenders, but they do request some "proof" that the borrower will be able to repay the loan.)

5. What is the true annual interest rate if an individual

    (a) borrows $5,000 now and repays it with a single payment of $6,000 in two months?

    (b) borrows $10,000 now and repays it with a single payment of $13,000 in four months?

6. Payday lenders sometimes charge up to $30 per $100 loaned. Complete this extension of Valarie's spreadsheet table:

Amount Borrowed	Loan Fee	True Rate for 2-Week Period	Nominal Annual Rate $i^{(26)}$	True Annual Rate
$100	$21			
$100	$22			
$100	$23			
$100	$24			
$100	$25			
$100	$26			
$100	$27			
$100	$28			
$100	$29			
$100	$30			

$$$
$$$

Session #35
WAYNE'S PROJECT: START TO SAVE EARLY

> *I finally know what distinguishes man from other beasts: financial worries.*
> -Jules Renard

Wayne's randomly-assigned project was to illustrate that the earlier one starts saving, the better. This command was understandably a bit vague to Wayne. He spent a bit of time discussing the assignment with Herkimer. As would have been true with any one of the ten intelligent members of the Stat Pack, Wayne was quick to understand the point of the project after the discussion.

When presenting his project, Wayne explained that he had constructed an interactive spreadsheet to illustrate the advantages of saving early. On the spreadsheet he was about to display, he said that the values one could input were a yearly deposit and a true annual interest rate. On the specific displayed sheet, the annual deposits were $5,000 and the true annual interest rate was 6%. Wayne described three different scenario situations:

SCENARIO #1:
Starting at age 20, individual A makes 11 annual deposits of $5,000. His last deposit is at age 30. He lets the money accumulate at 6% until age 65. At that age, he has made a total of $55,000 in deposits, has an accumulation of $575,367 and has earned $520,367 in interest.

SCENARIO #2:
Starting at age 31, individual B makes 34 annual deposits of $5,000. His last deposit is at age 64. He lets the money accumulate at 6% until age 65. At that age, he has made a total of $170,000 in deposits, has an accumulation of $552,174 and has earned $382,174 in interest.

SCENARIO #3:
Starting at age 20, individual C makes 45 annual deposits of $5,000. His last deposit is at age 64. He lets the money accumulate at 6% until age 65. At that age, he has made a total of $225,000 in deposits, has an accumulation of $1,127,541 and has earned $902,541 in interest.

Scenarios 1 and 2 were of particular interest. Wayne pointed out that the late-starting individual B invested much more than individual A, but came away with a smaller accumulation at age 65. He then displayed his spreadsheet outline displaying the year-by-year accumulation of the three investment scenarios:

189

Spreadsheet by WAYNE							
TOPIC: THE ADVANTAGES TO STARTING EARLY IN SAVING MONEY							
$5,000	<-- Yearly Deposit (INPUT)						
6%	<--Annual Interest Rate (INPUT)						
		SCENARIO #1		SCENARIO #2		SCENARIO #3	
Deposit #	Age	Deposit	Accum.	Deposit	Accum.	Deposit	Accum.
1	20	$5,000	$5,000			$5,000	$5,000
2	21	$5,000	$10,300			$5,000	$10,300
3	22	$5,000	$15,918			$5,000	$15,918
4	23	$5,000	$21,873			$5,000	$21,873
5	24	$5,000	$28,185			$5,000	$28,185
6	25	$5,000	$34,877			$5,000	$34,877
7	26	$5,000	$41,969			$5,000	$41,969
8	27	$5,000	$49,487			$5,000	$49,487
9	28	$5,000	$57,457			$5,000	$57,457
10	29	$5,000	$65,904			$5,000	$65,904
11	30	$5,000	$74,858			$5,000	$74,858
12	31		$79,350	$5,000	$5,000	$5,000	$84,350
13	32		$84,111	$5,000	$10,300	$5,000	$94,411
14	33		$89,157	$5,000	$15,918	$5,000	$105,075
15	34		$94,507	$5,000	$21,873	$5,000	$116,380
16	35		$100,177	$5,000	$28,185	$5,000	$128,363
17	36		$106,188	$5,000	$34,877	$5,000	$141,064
18	37		$112,559	$5,000	$41,969	$5,000	$154,528
19	38		$119,313	$5,000	$49,487	$5,000	$168,800
20	39		$126,471	$5,000	$57,457	$5,000	$183,928
21	40		$134,060	$5,000	$65,904	$5,000	$199,964
22	41		$142,103	$5,000	$74,858	$5,000	$216,961
23	42		$150,629	$5,000	$84,350	$5,000	$234,979
24	43		$159,667	$5,000	$94,411	$5,000	$254,078
25	44		$169,247	$5,000	$105,075	$5,000	$274,323
26	45		$179,402	$5,000	$116,380	$5,000	$295,782
27	46		$190,166	$5,000	$128,363	$5,000	$318,529
28	47		$201,576	$5,000	$141,064	$5,000	$342,641
29	48		$213,671	$5,000	$154,528	$5,000	$368,199
30	49		$226,491	$5,000	$168,800	$5,000	$395,291
31	50		$240,080	$5,000	$183,928	$5,000	$424,008
32	51		$254,485	$5,000	$199,964	$5,000	$454,449
33	52		$269,754	$5,000	$216,961	$5,000	$486,716
34	53		$285,940	$5,000	$234,979	$5,000	$520,919
35	54		$303,096	$5,000	$254,078	$5,000	$557,174
36	55		$321,282	$5,000	$274,323	$5,000	$595,604
37	56		$340,559	$5,000	$295,782	$5,000	$636,341
38	57		$360,992	$5,000	$318,529	$5,000	$679,521
39	58		$382,652	$5,000	$342,641	$5,000	$725,292
40	59		$405,611	$5,000	$368,199	$5,000	$773,810
41	60		$429,947	$5,000	$395,291	$5,000	$825,238
42	61		$455,744	$5,000	$424,008	$5,000	$879,753
43	62		$483,089	$5,000	$454,449	$5,000	$937,538
44	63		$512,074	$5,000	$486,716	$5,000	$998,790
45	64		$542,799	$5,000	$520,919	$5,000	$1,063,718
46	65		$575,367		$552,174		$1,127,541
Total Accumulation			$575,367		$552,174		$1,127,541
Total All Payments			$55,000		$170,000		$225,000
Total Interest earned			$520,367		$382,174		$902,541

190

Wayne went on to demonstrate that he could use what the Pack had learned in earlier sessions to verify the spreadsheet totals. He emphasized that scenario #1 involved accumulating a series of payments for 11 years, obtaining $79,350 at age 31, and then accumulating this single amount for 34 years. He put this handwritten transparency on the overhead:

SCENARIO #1:

$$\$5000(1.06)\left[\frac{(1.06)^{11}-1}{.06}\right](1.06)^{34}$$

$$= \$575,367.$$

SCENARIO #2:

$$\$5000(1.06)\left[\frac{(1.06)^{34}-1}{.06}\right]$$

$$= \$552,174.$$

SCENARIO #3:

$$\$5000(1.06)\left[\frac{(1.06)^{45}-1}{.06}\right]$$

$$= \$1,127,541.$$

Herkimer praised Wayne for a good presentation that emphasized a vital point. "The power of compounding is amazing," he said. "Wayne has clearly demonstrated the advantages of beginning to save at an early age, particularly if one is thinking about saving for retirement. And, remember that spreadsheet computations are only as accurate as the individual who created the spreadsheet. It's nice to have those financial formulas we used many sessions ago to verify spreadsheet computations."

HERKIMER'S FASCINATING FINANCIAL FACTS:

When was paper money first printed in the United States?

The answer might surprise you. It was in 1862. The U.S. Department of the Treasury issued paper money to make up for a shortage of coins and to finance the Civil War. The uncertainty caused by the war caused people to start hoarding them. Values of everyday items fluctuated drastically, but because coins were made of gold and silver their value didn't fluctuate much. People tended to hold onto them rather than buy items that might lose value over a short period of time.

ACTIVITY SET FOR SESSION #35

1. **ALCAL** activity: Use your calculator to check the Wayne's handwritten computation for each of the three scenarios he presented:

   (a) Scenario #1:

To be evaluated	$5000(1.06)[(1.06^{11}-1)/.06])1.06)$
Calculator entry	5000*1.06*(1.06^11-1)/.06*1.06^34
Number represented	

   (b) Scenario #2:

To be evaluated	$5000(1.06)(1.06^{34}-1)/.06$
Calculator entry	5000*1.06*(1.06^34-1)/.06
Number represented	

   (c) Scenario #3:

To be evaluated	$5000(1.06)(1.06^{45}-1)/.06$
Calculator entry	5000*1.06*(1.06^45-1)/.06
Number represented	

2. REFLECTION & COMMENT activity:

   Not only does it pay to save, but if you start sooner you can take advantage of the power of compounding. For example, your deposits earn interest and so does you reinvested interest. This is letting your money work for you. The sooner you start saving for retirement, the more you will have when you retire. And the sooner you start saving for retirement, the sooner you will be able to retire.

   (Alex Gallego, *What Are the Advantages of Saving Sooner?*, northender.com)

3. Consider the three scenarios presented in the vignette. Find the respective totals at age 65 if the true annual interest rate is

    (a) 5%   (b) 7%   (c) 10%.

4. Plan #1 has an individual starting at age 20 and making 20 annual deposits of $10,000 the last deposit at age 39. This money will be left on deposit until age 65. Plan #2 has an individual depositing a single sum of $300,000 at age 40 and letting this money accumulate to age 65. If all payments earn a true annual rate i, express in terms of i the accumulated amounts at age 65.

5. Referencing activity #2, find the accumulated amounts for Plan #1 and Plan #2 for these true annual rates:

    (a) i = 5%   (b) i = 7%   (c) i = 10%

6. Create other investment scenarios that illustrate the advantages of saving early.

$$
$$$

> *Money is better than poverty, if only for financial reasons.*
> -Woody Allen

Darren began his presentation on continuous interest by noting that if one invested $1 at a true annual rate of 100%, he would have $1(1 + 1) = $2 after one year. "Now," he asked, "what if the 100% is a nominal rate? What if it is compounded monthly? Or daily? Or 1000 times a year?"

The first reaction of Pack members was to reflect the thought that the $1 would increase indefinitely as the number of compounding times increased. "My research indicates that this is what a lot of people think," said Darren. He reminded the Pack that if an annual interest rate i is compounded n times a year, then $1 would accumulate to

$$\$1(1 + i/n)^n$$

after one year. Darren note that 100% = 1, so if an annual rate of 100% is compounded n times a year, then $1 would accumulate to

$$\$1(1 + 1/n)^n$$

after 1 year.

Darren asked the Pack to use their calculators to "test out" the value of this expression as the value of n got larger and larger. They did this and the group was somewhat surprised at the results. For instance, they found that if $i^{(365)} = 100\%$, then $\$1(1 + 1/365)^{365} = \$2.71$ when rounded to the nearest penny. Darren then put this transparency on the overhead:

Annual rate of 100%	x = number of times compounded per year	Value of $(1 + 1/x)^x$	Value of e (9 decimal places)
True rate	1	2.000000000	2.718281828
Compounded semiannually	2	2.250000000	2.718281828
Compounded quarterly	4	2.441406250	2.718281828
Compounded monthly	12	2.613035290	2.718281828
Compounded daily	365	2.714567482	2.718281828
Compounded hourly	8,760	2.718126692	2.718281828
Compounded every minute	525,600	2.718279243	2.718281828
Compounded every second	31,536,000	2.718281781	2.718281828

All Pack members had at least pre-calculus in their math backgrounds. Initially puzzled by the fourth column on Darren's table, they soon realized they were looking at the value of e, the base of natural logarithms, expressed to nine decimal places.

194

Darren then reminded the Pack that if a deposit D accumulates at an annual rate i that is compounded N times a year, then after 1 year the accumulation is

$$D(1 + i/N)^N$$

He then put this hand-written transparency on the overhead, noting that an understanding of some basic algebra and limits was important in the derivation of e from the financial expression that allowed one to accumulate money at nominal annual rates.

$$\lim_{N \to \infty} D\left(1 + \frac{i}{N}\right)^N = D \lim_{N \to \infty} \left(1 + \frac{i}{N}\right)^N$$

$$= D \lim_{N \to \infty} \left(1 + \frac{1}{\frac{N}{i}}\right)^N$$

$$= D \lim_{N \to \infty} \left[\left(1 + \frac{1}{\frac{N}{i}}\right)^{\frac{N}{i}}\right]^i$$

As $N \to \infty$, $\frac{N}{i} \to \infty$ also.

Hence

$$D \lim_{N \to \infty} \left[\left(1 + \frac{1}{\frac{N}{i}}\right)^{\frac{N}{i}}\right]^i$$

$$= De^i.$$

Example: If $\$1,000$ earns 8% per annum compounded continuously, then the accumulation after one year is

$$\$1,000\, e^{.08} = \$1,083.29.$$

Darren then displayed an interactive spreadsheet he constructed that accumulated beginning-of-year deposits for 20 years. The sheet allowed the user to input both a deposit and an annual rate. The purpose was to illustrate the effects of compounding.

Spreadsheet by DARREN						
Topic: Accumulation of beginning-of-year deposits for 20 years at a specified						
rate compounded numerous times per year.						
$100,000	<---INPUT beginning-of-year deposit					
8%	<---INPUT annual interest rate					
	Number of times compounded per year					
	1	2	4	12	365	Continuously
Year						
0	$100,000	$100,000	$100,000	$100,000	$100,000	$100,000
1	$208,000	$208,160	$208,243	$208,300	$208,328	$208,329
2	$324,640	$325,146	$325,409	$325,589	$325,677	$325,680
3	$450,611	$451,678	$452,233	$452,612	$452,798	$452,805
4	$586,660	$588,535	$589,512	$590,179	$590,506	$590,517
5	$733,593	$736,559	$738,107	$739,164	$739,682	$739,700
6	$892,280	$896,662	$898,950	$900,514	$901,281	$901,307
7	$1,063,663	$1,069,830	$1,073,053	$1,075,256	$1,076,338	$1,076,375
8	$1,248,756	$1,257,128	$1,261,507	$1,264,502	$1,265,972	$1,266,023
9	$1,448,656	$1,459,710	$1,465,496	$1,469,455	$1,471,400	$1,471,466
10	$1,664,549	$1,678,822	$1,686,300	$1,691,419	$1,693,934	$1,694,020
11	$1,897,713	$1,915,814	$1,925,305	$1,931,806	$1,935,001	$1,935,110
12	$2,149,530	$2,172,144	$2,184,012	$2,192,145	$2,196,143	$2,196,280
13	$2,421,492	$2,449,391	$2,464,045	$2,474,092	$2,479,033	$2,479,201
14	$2,715,211	$2,749,262	$2,767,161	$2,779,440	$2,785,480	$2,785,687
15	$3,032,428	$3,073,601	$3,095,264	$3,110,132	$3,117,448	$3,117,699
16	$3,375,023	$3,424,407	$3,450,414	$3,468,272	$3,477,062	$3,477,363
17	$3,745,024	$3,803,839	$3,834,839	$3,856,136	$3,866,623	$3,866,982
18	$4,144,626	$4,214,232	$4,250,953	$4,276,194	$4,288,626	$4,289,051
19	$4,576,196	$4,658,114	$4,701,368	$4,731,116	$4,745,773	$4,746,274
20	$5,042,292	$5,138,216	$5,188,912	$5,223,796	$5,240,989	$5,241,577

The Pack was quick to notice that there really wasn't much difference between daily compounding and continuous compounding. Darren indicated that his research showed this was generally true. He also noted that continuous compounding allows one to work with less complicated expressions. For instance, if an annual rate i is compounded daily, then $1 grows to $1(1 + i/365)^{365}$ after one year. If the rate is compounded continuously, the amount at the end of one year is $1e^{i}$.

Darren was praised by Herkimer and the Pack for a job well done.

HERKIMER'S FASCINATING FINANCIAL FACTS:

Many people save two dollar bills thinking that they are rare and therefore valuable. They were first issued in 1862 and those from the 19th century have some value, but the value of virtually any two dollar bill is simply $2. The modern two dollar bill has President Thomas Jefferson on the front and the back in an engraving of John Turnbull's painting, *The Signing of the Declaration of Independence*. Due to lack of space, 5 of the 47 men in Turnbull's original painting were not included in the engraving. Modern cash registers don't have a slot for a $2 bill.

ACTIVITY SET FOR SESSION #36

1. **ALCAL** activity: Use your calculator to check the end-of-first year values displayed on Darren's spreadsheet for the indicated rates:

(a) 8% compounded quarterly:

To be evaluated	$100000+10000(1+.08/4)^4$
Calculator entry	100000+100000*(1+.08/4)^4
Number represented	

(b) 8% compounded daily:

To be evaluated	$100000+10000(1+.08/365)^{365}$
Calculator entry	100000+100000*(1+.08/365)^365
Number represented	

(c) 8% compounded continuously:

To be evaluated	$100000+100000e^{.08}$
Calculator entry	100000+100000*e^(.08)
Number represented	

2. REFLECTION & COMMENT activity:

*Back when Elvis was King and computers were scarce banks used to compound interest quarterly. That meant that four times a year they would have an "interest day" when everybody's balance got bumped up by one-fourth of the going interest rate … and bank employees would have to work late, going home all sweaty and covered with ink. If you held an account in those days, every year your balance would increase by a factor of $(1+r/4)^4$. Today it's possible to compound interest monthly, daily, and in the limiting case, **continuously**, meaning that your balance grows by a small amount every instant.*

(www.moneychimp.com, June 2009))

197

3. Find the accumulation of a single deposit of $100,000 at the end of 30 years at each annual rate:

   (a) a true rate of 10%   (b) $i^{(4)} = 10\%$ (c) $i^{(12)} = 10\%$

   (d) $i^{(365)} = 10\%$ (e) Compounded continuously at 10%

4. A deposit of $1000 is left to accumulate for 50 years. If the deposit earns a guaranteed annual rate of 10%, what will be the accumulation after 50 years if the rate is

   (a) a true rate of 10%?   (b) compounded continuously at 10%?

5. How long would it take a deposit of $1 to accumulate to $5 if the annual interest rate is

   (a) a true rate of 7%?   (b) compounded continuously at 7%?

6. Beginning-of-year deposits of $10,000 are made for 5 years as indicated in the table. The deposits earn an annual rate of 10%. Complete the table by filling in the appropriate amounts to the nearest dollar:

Year	Deposit	Accumulation at true rate of 10%	Accumulation at 10% compounded continuously
0	$10,000	$10,000	$10,000
1	$10,000		
2	$10,000		
3	$10,000		
4	$10,000		
5			

$$
$$

Session #37
FRANCES' PROJECT: ADVANTAGE OF EXTRA MORTGAGE PAYMENTS

> *I don't mind going back to daylight saving time. With inflation, the hour will be the only thing I've saved all year.*
> -Victor Borge

"You generally don't want to pay more than you owe," began Frances, "but this may be beneficial when it comes to mortgages. We've learned that the interest portion of any payment is determined by what you owe after the previous payment. So, anything that reduces what you owe is really beneficial to you."

While searching the internet with phrases like ***advantages of making extra payments on a mortgage***, Frances found a number of sites that provided a built-in calculator to illustrate what she had just said. One particular site allowed the user to do the following:

INPUT the amount of the mortgage.

INPUT the mortgage time period in terms of years.

INPUT the mortgage interest rate (an annual rate compounded monthly).

INPUT the extra amount paid each month.

The user would then be able to view the following output for the mortgage displayed:

INPUT Amount of mortgage	$400,000
INPUT Mortgage term (years)	30
INPUT Mortgage rate	7.5%
INPUT Extra amount paid each month	$100
Monthly payment	$2,797
Monthly payment with extra amount	$2,897
This shortens mortgage by	3 years, 4 months
Savings	$82,651

Frances told the Pack that she thought these internet sites were neat, but they simply returned values that one had to take on faith. She reminded the group that they had actually constructed complete mortgage payment schedules on spreadsheets in previous sessions. She decided to use these spreadsheets to check on the internet output she obtained.

Now a 30-year mortgage would have 360 lines of payment breakdowns. So Frances used the HIDE ROW spreadsheet command to show only the relevant portions of the schedule. She said she would show the row and column heading on her spreadsheets so that one could actually see the numbers of the rows hidden. Here is what she put on the overhead relating to the previously displayed internet output:

	A	B	C	D	E	F	G	H	I
1	**30-YEAR MORTGAGE PAYMENT SCHEDULE**								
2									
3									
4	**MORTGAGE SPREADSHEET...constructed by Herkimer's Stat Pack**								
5	User inputs LOAN AMOUNT, INTEREST RATE, and NUMBER OF YEARS (TERM OF MORTGAGE).								
6									
7	**$400,000**	<---INPUT LOAN AMOUNT							
8	**7.50%**	<---INPUT INTEREST RATE (Annual Rate Compounded Monthly)							
9	**30**	<---INPUT NUMBER OF YEARS (TERM OF MORTGAGE)							
10									
11	**$2,796.86**	<----MONTHLY PAYMENT (Calculated)							
12									
13	Month #	**Payment**	**Interest**	**Principal**	**Princ. Outstand.**		% Interest in payment		
14	0				$400,000.00				
15	1	$2,796.86	$2,500.00	$296.86	$399,703.14		89.39%		
16	2	$2,796.86	$2,498.14	$298.71	$399,404.43		89.32%		
17	3	$2,796.86	$2,496.28	$300.58	$399,103.85		89.25%		
371								**Hidden rows**	
372	358	$2,796.86	$51.79	$2,745.07	$5,541.71		1.85%		
373	359	$2,796.86	$34.64	$2,762.22	$2,779.49		1.24%		
374	360	$2,796.86	$17.37	$2,779.49	$0.00		0.62%	30 years	
375									
376	**TOTALS**	**$1,006,868.89**	**$606,868.89**	**$400,000.00**	<---THESE ARE		**30**	YEAR TOTALS	

Frances adjusted the spreadsheet to reflect the extra payment of $100.

	A	B	C	D	E	F	G	H	I
3									
4	**MORTGAGE SPREADSHEET...constructed by Herkimer's Stat Pack**								
5	User inputs LOAN AMOUNT, INTEREST RATE, and NUMBER OF YEARS (TERM OF MORTGAGE).								
6									
7	**$400,000**	<---INPUT LOAN AMOUNT							
8	**7.50%**	<---INPUT INTEREST RATE (Annual Rate Compounded Monthly)							
9	**30**	<---INPUT NUMBER OF YEARS (TERM OF MORTGAGE)							
10									
11	**$2,796.86**	<----MONTHLY PAYMENT (Calculated)							
12									
13		**With extra $100**							
14	Month #	**Payment**	**Interest**	**Principal**	**Princ. Outstand.**		% Interest in payment		
15	0				$400,000.00				
16	1	$2,896.86	$2,500.00	$396.86	$399,603.14		86.30%		
17	2	$2,896.86	$2,497.52	$399.34	$399,203.80		86.21%		
18	3	$2,896.86	$2,495.02	$401.84	$398,801.96		86.13%		
332								**Hidden rows**	
333	318	$2,896.86	$36.60	$2,860.26	$2,995.72		1.26%		
334	319	$2,896.86	$18.72	$2,878.14	$117.59		0.65%		
335	320	$118.32	$0.73	$117.59	$0.00		0.62%		
376									
377	**TOTALS**	**$924,216.66**	**$524,216.66**	**$400,000.00**					
378									

As the second spreadsheet indicates, the extra $100 reduced the time to 320 months. This is 40 fewer monthly payments, representing a time reduction of 3 years and 4 months. The indicated savings: $1,006,868.89 - $924,216.66 = $82,652.23. The internet output was clearly "on the money" with the small difference of about $1 due to rounding to the nearest dollar on the internet output.

Frances produced another example using the internet calculator. This one involved a 20 year $220,000 mortgage at 6.8% with an extra payment of $200 each month. The internet output was as follows:

INPUT Amount of mortgage	**$220,000**
INPUT Mortgage term (years)	**20**
INPUT Mortgage rate	**6.8%**
INPUT Extra amount paid each month	**$200**
Monthly payment	$1,679
Monthly payment with extra amount	$1,879
This shortens mortgage by	3 years, 11 months
Savings	$40,948

She then demonstrated the output obtained from spreadsheets previously constructed by Pack members relating to 20 year mortgages:

	A	B	C	D	E	F	G	H	I
1	MORTGAGE SPREADSHEET...constructed by Herkimer's Stat Pack								
2	User inputs LOAN AMOUNT, INTEREST RATE, and NUMBER OF YEARS (TERM OF MORTGAGE).								
3									
4	$220,000	<---INPUT LOAN AMOUNT							
5	6.80%	<---INPUT INTEREST RATE (Annual Rate Compounded Monthly)							
6	20	<---INPUT NUMBER OF YEARS (TERM OF MORTGAGE)							
7									
8	$1,679.35	<----MONTHLY PAYMENT (Calculated)							
9									
10	Month #	Payment	Interest	Principal	Princ. Outstand.		% Interest in payment		
11	0				$220,000.00				
12	1	$1,679.35	$1,246.67	$432.68	$219,567.32		74.24%		
13	2	$1,679.35	$1,244.21	$435.13	$219,132.19		74.09%		
14	3	$1,679.35	$1,241.75	$437.60	$218,694.59		73.94%		
15								hidden rows	
249	238	$1,679.35	$28.23	$1,651.12	$3,330.36		1.68%		
250	239	$1,679.35	$18.87	$1,660.47	$1,669.88		1.12%		
251	240	$1,679.35	$9.46	$1,669.88	$0.00		0.56%	20 years	
372									
373	TOTALS	$403,043.27	$183,043.27	$220,000.00	<---THESE ARE		20	YEAR TOTALS	

As she had done in the previous example, Frances adjusted the payment to include the extra $200 per month.

	A	B	C	D	E	F	G	H	I
3									
4	$220,000	<---INPUT LOAN AMOUNT							
5	6.80%	<---INPUT INTEREST RATE (Annual Rate Compounded Monthly)							
6	20	<---INPUT NUMBER OF YEARS (TERM OF MORTGAGE)							
7									
8	$1,679.35	<----MONTHLY PAYMENT (Calculated)							
9									
10		With extra $200							
11	Month #	Payment	Interest	Principal	Princ. Outstand.		% Interest in payment		
12	0				$220,000.00				
13	1	$1,879.35	$1,246.67	$632.68	$219,367.32		66.33%		
14	2	$1,879.35	$1,243.08	$636.27	$218,731.05		66.14%		
15	3	$1,879.35	$1,239.48	$639.87	$218,091.17		65.95%		
201								hidden rows	
202	190	$1,879.35	$38.57	$1,840.78	$4,964.95		2.05%		
203	191	$1,879.35	$28.13	$1,851.22	$3,113.73		1.50%		
204	192	$1,879.35	$17.64	$1,861.71	$1,252.03		0.94%	16 years	
205	193	$1,259.12	$7.09	$1,252.03	$0.00		0.56%		
374	TOTALS	$362,094.32	$142,094.32	$220,000.00					
375									

The Pack-produced spreadsheet indicated that the mortgage would be paid off in 193 months. This is 16 years and 1 month. This shortened the mortgage by 3 years and 11 months, as indicated in the internet output. The total amount saved would be $403,043.27 - $362,094.32 = $40,948.95. This "jived" with the output from the internet source.

Frances repeated an important point: "The interest in any payment is calculated on the amount you still owe on the mortgage, so whatever you can do to reduce the principal is beneficial." She went on to say that there are a variety of ways one could make extra payments. For instance, one could just make an additional payment each year. Or, if a large amount of money suddenly became available, this could be added to a payment and shorten the time span of a mortgage.

Frances got her point across to the Pack. And, they were enthusiastic about the fact that they have constructed financial spreadsheets that could be used to verify output from sources in the media.

ACTIVITY SET FOR SESSION #37

1. **ALCAL** activity: Use your calculator to check the monthly payments on these mortgages featured in Frances' spreadsheets:

(a) A 30-year $400,000 mortgage at 7.50%:

To be evaluated	$400000(.075/12)/[1-(1+.075/12)^{-360}]$
Calculator entry	400000*.075/12/(1-(1+.075/12)^360)
Number represented	

(b) A 20-year $220,000 mortgage at 6.8%:

To be evaluated	$220000(.068/12)/[1-(1+.068/12)^{-240}]$
Calculator entry	220000*.068/12/(1-(1+.068/12)^240)
Number represented	

2. REFLECTION & COMMENT activity:

The main advantage of making extra payments is that by reducing your principal loan balance, you are also reducing the amount of interest you pay over time. Making extra payments on a regular basis can help substantially reduce the total cost of your loan. Note, some lenders may impose an "early payment" penalty, so check with your lender before choosing a loan if you are interested in making extra payments.

(JP Morgan Chase & Co., *How Advantageous are Extra Payments ?*, June 2009))

3. Research the internet for sources relating to the advantages of making extra payments on a mortgage. Find a source that allows the user to use a built-in calculator to indicate the savings resulting from extra payments.

Questions 4 - 6 relate to a 15-year $200,000 mortgage with a 7.2% annual interest rate.

4. Calculate the monthly payment for this mortgage and the total amount paid over the 15 year period.

5. Assume you make one extra payment a year on this mortgage and that you do this by paying twice the amount owed on payments 1, 13, 25, 37, 49, etc. How long would it take you to pay off the mortgage? Calculate the total amount you would pay and the amount saved by making the extra payments.

6. Assume that you pay an extra $10,000 in the 13th payment on this mortgage. How long would it take you to pay off the mortgage? Calculate the total amount you would pay and the amount saved by making the extra payment.

----

7. Consider credit card debt and investigate the advantages of paying as much of what you owe as soon as possible. (Given that credit card rates are high compared to conventional loans, is it really surprising that wise financial advice usually suggests ridding yourself of credit card debt before addressing other financial problems?)

$$$
$$

Session #38
CAROLYN'S PROJECT: REFINANCING

> *Finance is the art of passing currency from hand to hand until it finally disappears.*
> -Robert W. Sarnoff

Some Pack members said they had heard their parents discuss refinancing their present mortgage. Carolyn said that her research indicated that if interest rates have dropped, it might be beneficial to refinance the loan that represents the mortgage. This could lead to lower monthly payments and perhaps save thousands of dollars over the course of many years. "I learned that you have to be careful doing this," she said, "because there are costs associated with refinancing. These costs can be about 3% of the loan itself, and perhaps as much as 6%. Some articles I read said that if you can get an interest rate that is at least 2% less than the one you have, it might well be worth exploring the possibility of refinancing."

Prior to beginning her presentation, Carolyn had written some material on the chalkboard. She wanted the Pack to have a review of mortgage mathematics. She used a 30-year $300,000 mortgage with an 8% rate to jog memories and to remind the group of the financial mathematics power they had gained during their work with Herkimer.

Thirty year $300,000 mortgage at 8%.

Monthly payment

$$= \frac{(\$300,000)(.08/12)}{1 - \left(1 + \frac{.08}{12}\right)^{-360}} = \$2,201.29.$$

Total payments = $(\$2,201.29)(360) = \$792,464.40.$

Carolyn told the Pack that she would show them some refinancing possibilities on this mortgage after it had been in existence for two years. She used the top portion of the mortgage spreadsheet that the Pack had developed many sessions ago.

205

30-YEAR MORTGAGE PAYMENT SCHEDULE

MORTGAGE SPREADSHEET…constructed by Herkimer's Stat Pack
User inputs LOAN AMOUNT, INTEREST RATE, and NUMBER OF YEARS (TERM OF MORTGAGE).

$300,000	<---INPUT LOAN AMOUNT
8.00%	<---INPUT INTEREST RATE (Annual Rate Compounded Monthly)
30	<---INPUT NUMBER OF YEARS (TERM OF MORTGAGE)

$2,201.29	<----MONTHLY PAYMENT (Calculated)

Month #	Payment	Interest	Principal	Princ. Outstand.	% Interest in payment	
0				$300,000.00		
1	$2,201.29	$2,000.00	$201.29	$299,798.71	90.86%	
2	$2,201.29	$1,998.66	$202.64	$299,596.07	90.79%	
3	$2,201.29	$1,997.31	$203.99	$299,392.08	90.73%	
4	$2,201.29	$1,995.95	$205.35	$299,186.74	90.67%	
5	$2,201.29	$1,994.58	$206.72	$298,980.02	90.61%	
6	$2,201.29	$1,993.20	$208.09	$298,771.93	90.55%	
7	$2,201.29	$1,991.81	$209.48	$298,562.45	90.48%	
8	$2,201.29	$1,990.42	$210.88	$298,351.57	90.42%	
9	$2,201.29	$1,989.01	$212.28	$298,139.29	90.36%	
10	$2,201.29	$1,987.60	$213.70	$297,925.59	90.29%	
11	$2,201.29	$1,986.17	$215.12	$297,710.47	90.23%	
12	$2,201.29	$1,984.74	$216.56	$297,493.91	90.16%	1 year
13	$2,201.29	$1,983.29	$218.00	$297,275.91	90.10%	
14	$2,201.29	$1,981.84	$219.45	$297,056.45	90.03%	
15	$2,201.29	$1,980.38	$220.92	$296,835.54	89.96%	
16	$2,201.29	$1,978.90	$222.39	$296,613.15	89.90%	
17	$2,201.29	$1,977.42	$223.87	$296,389.27	89.83%	
18	$2,201.29	$1,975.93	$225.37	$296,163.91	89.76%	
19	$2,201.29	$1,974.43	$226.87	$295,937.04	89.69%	
20	$2,201.29	$1,972.91	$228.38	$295,708.66	89.63%	
21	$2,201.29	$1,971.39	$229.90	$295,478.76	89.56%	
22	$2,201.29	$1,969.86	$231.44	$295,247.32	89.49%	
23	$2,201.29	$1,968.32	$232.98	$295,014.34	89.42%	
24	$2,201.29	$1,966.76	$234.53	**$294,779.81**	89.35%	2 years

"As you can see," said Carolyn, "at the end of two years the outstanding principal on this mortgage is about $295,000." She went on to indicate that the remaining 336 payments of $2,201.29 totaled to $739,633.44.

Carolyn indicated that she was going to display some refinancing possibilities for the $295,000. "Nothing is automatic here," she said. "People wishing to finance have to qualify for the new loan. And, the refinancing schedule might look good, but keep in mind there are costs associated with a new loan." She indicated that these are usually called *closing costs* and if they amounted to 3% of the loan, they would be (0.03)($295,000) = $8,850. She then put this neatly-prepared transparency on the overhead and asked her classmates to check the displayed totals. It compared the original mortgage (now expressed as a 28-year loan since 2 years of payments have been made). She also assumed closing cost for refinancing would be 3% of the loan amount.

	ORIGINAL MORTGAGE				
Loan Amount	$294,779.81	$295,000.00	$295,000.00	$295,000.00	$295,000.00
Loan Rate	8%	6%	6%	6%	6%
Term (Years)	28	30	25	20	15
Monthly Payment	$2,201.29	$1768.67	$1,900.69	$2,113.47	$2,489.38
Total Payments	$739,633.44	$636,721.20	$570,207.00	$507,232.08	$448,088.40
Closing costs	$0.00	$8,850.00	$8,850.00	$8,850.00	$8,850.00
Total Cost	$739,633.44	$645,571.20	$579,057.00	$516,082.08	$456,938.40

Overall it did appear that refinancing with an interest rate 2% lower than an existing one could be beneficial. Also, lessening the term of the loan could actually increase monthly payments but result in an overall total cost significantly smaller than the original total. Carolyn indicated once again that someone looking into refinancing definitely has to consider the closing costs since they usually must be paid before the loan becomes active. Pack members used their calculators to verify that Carolyn's figures were indeed correct.

"Even a difference of 1% produces some interesting results," continued Carolyn. She put this transparency on the overhead displaying refinancing possibilities with just a 1% difference in rates:

	ORIGINAL MORTGAGE				
Loan Amount	$294,779.81	$295,000.00	$295,000.00	$295,000.00	$295,000.00
Loan Rate	8%	7%	7%	7%	7%
Term (Years)	28	30	25	20	15
Monthly Payment	$2,201.29	$1,962.64	$2,085.00	$2,287.13	$2,651.54
Total Payments	$739,633.44	$706,550.40	$625,500.00	$548,911.20	$477,277.20
Closing costs	$0.00	$8,850.00	$8,850.00	$8,850.00	$8,850.00
Total Cost	$739,633.44	$715,400.40	$634,350.00	$557,761.20	$486,127.20

"A nice presentation," said Herkimer. "I hope you young folks realize that when you get a bit older you won't need to hire someone to prepare financial statements for you relating to mortgage possibilities. You have the MATH POWER to do it yourselves."

HERKIMER'S FASCINATING FINANCIAL FACTS:

Defacement of currency is a purposeful act that makes it unfit for circulation. This act is punishable by a fine or imprisonment for not more than six months. Basically, anyone who cuts, mutilates, perforates, disfigures, or cements together currency to make it unfit for use is guilty of defacement. The act of defacement is defined in Title 18, Section 333 of the United States Code that is enforced by the United States Secret Service.

ACTIVITY SET FOR SESSION #38

1. **ALCAL** activity: Use your calculator to check the computations displayed on the 30-year $300,000 mortgage at 8% used by Carolyn in her presentation. Keep in mind that in using the payment $2,201.29 results will differ slightly due to rounding. The spreadsheet computation used more than two decimal places although only two are displayed on the sheet itself. The calculator check makes use of the fact that the principal outstanding at any time is the present value of the remaining payments:

(a) The principal outstanding after 1 year:

To be evaluated	$2201.29[1-(1+.08/12)^{-348}]/(.08/12)$
Calculator entry	2201.29*(1-(1+.08/12)^348)/(.08/12)
Number represented	

(b) The principal outstanding after 2 years:

To be evaluated	$2201.29[1-(1+.08/12)^{-336}]/(.08/12)$
Calculator entry	2201.29*(1-(1+.08/12)^336)/(.08/12)
Number represented	

2. REFLECTION & COMMENT activity:

In the case of refinancing, most people only care about the immediate, short-term savings, failing to see the real costs they will incur over the long run.

(*Money Mastery*, by Williams, Jeppson, and Botkin)

3. As Carolyn indicated in the vignette, there are closing costs related to financing. Below is a list of fees one might well encounter in refinancing. Research these terms to find out what they represent:

Loan origination fee	Discount points	Application fee	Appraisal fee	Title examination fee
Title insurance for lender	Land survey fee	Credit report fee	Documentation fee	Legal fees

4. Consider a 30-year $250,000 mortgage at 7.5%. Verify that the monthly payment is $1,748.04 and that the principal outstanding at the end of the 6th year is $233,194.35. Assume that after 6 years the borrower has the opportunity to refinance at 6% on a loan of $233,000. Assume that closing costs are 4% of the loan amount. Complete the following table illustrating various refinance options. (The original loan would be reduced to 24 years as indicated in the first column.)

	ORIGINAL MORTGAGE				
Loan Amount	$233,194.35	$233,000.00	$233,000.00	$233,000.00	$233,000.00
Loan Rate	7.5%	6%	6%	6%	6%
Term (Years)	24	30	25	20	15
Monthly Payment	$1,748.04				
Total Payments	$503,435.52				
Closing costs	$0.00				
Total Cost	$503,435.52				

5. Consider a 25-year $180,000 mortgage at 7%. Verify that the monthly payment is $1,272.20 and that the principal outstanding at the end of the 5th year is $164,091.87. Fill in the following table to display the results of refinancing $164,000 at the respective rates and terms. Assume that closing costs are 5% of the loan amount. (The original loan would be reduced to 20 years as indicated in the first column.)

	ORIGINAL MORTGAGE				
Loan Amount	$164,091.87	$164,000.00	$164,000.00	$164,000.00	$164,000.00
Loan Rate	7%	6%	5%	6%	5%
Term (Years)	20	20	20	15	15
Monthly Payment	$1,272.20				
Total Payments	$305,328.00				
Closing costs	$0.00				
Total Cost	$305,328.00				

$$
$$

> *Money was never a big motivation for me, except as a way to keep score. The real excitement is playing the game.*
> -Donald Trump

Roger reminded the Pack that a bond is really a loan and that the owner of a bond has an asset that has a value at any specific time. As coupons (interest) is paid, the value of the asset changes. A schedule displaying the value of a bond over a time period is called an *amortization schedule*. The value of a bond on a specific date is called its *book value*, or *amortized value*.

In their previous study of bonds Herkimer had the Pack work only with annual coupons. Many bonds issue semiannual coupons. Herkimer asked that Roger focus on bonds with semiannual coupons. With these bonds the coupon rate and yield rate are annual rates compounded semiannually.

Roger asked the Pack to consider a $10,000 bond with 6% coupons paid semiannually. Referencing previous financial lingo, the coupon rate was $i^{(2)} = 6\%$. Coupons of $300 were paid every 6 months. To demonstrate how bond amortization works, Roger had prepared a type-written transparency for the group to view to illustrate the computation involved in producing an amortization schedule. This showed schedules for this bond for a two-year period. He mentioned that this could be a two-year bond, or that one could simply be purchasing the final two years of an existing bond with a longer term that had been paying coupons over a period of many years.

Roger also jogged Pack memories by stating that if the coupon rate was less than the yield rate, one would pay less than the bond face value. His first example assumed a yield rate $i^{(2)} = 8\%$. The interest paid through the coupons was less than what was required, so one would "write up" the asset value of the bond after a coupon payment to compensate. In his second example, the coupon rate $i^{(2)} = 6\%$ is more than the yield rate $i^{(2)} = 4\%$. Hence, the coupon actually pays more interest than that which is required. In this case one would "write down" the book value after the payment of a coupon.

Roger placed this transparency on the overhead and gave the group time to study it:

If purchased to yield $i^{(2)} = 8\%$, then

purchase price = $300[1 - (1.04)^{-4}]/.04 + \$10,000/(1.04)^4 = \$9,630.01$

AMORTIZATION SCHEDULE:

Coupon #	Year #	A = Interest Required	C = Coupon	Book Value Change (A-C)	Book Value
0	0				**$9,637.01**
1	0.5	(.04)($9,637.01) = **$385.48**	**$300.00**	$385.48 - $300.00 = **$85.48**	$9,637.01 + $85.48 = **$9,722.49**
2	1	(.04)($9,722.49) = **$388.90**	**$300.00**	$388.90 - $300.00 = **$88.90**	$9,722.49 + $88.90 = **$9,811.39**
3	1.5	(.04)($9,811.39) = **$392.46**	**$300.00**	$392.46 - $300.00 = **$92.46**	$9,811.39 + $92.46 = **$9,903.85**
4	2	(.04)($9,903.85) = **$396.15**	**$300.00**	$396.15 - $300.00 = **$96.15**	$9,903.85 + $96.15 = **$10,000.00**

If purchased to yield $i^{(2)} = 4\%$, then

purchase price = $300[1 - (1.02)^{-4}]/.02 + \$10,000/(1.02)^4 = \$10,380.77$

AMORTIZATION SCHEDULE:

Coupon #	Year #	A = Interest Required	C = Coupon	Book Value Change (A-C)	Book Value
0	0				**$10,380.77**
1	0.5	(.02)($10,380.77) = **$207.62**	**$300.00**	$207.72 - $300.00 = **-$92.38**	$10,380.77+(-$92.38) = **$10,288.39**
2	1	(.02)($10,288.39) = **$205.77**	**$300.00**	$205.77 - $300.00 = **-$94.23**	$10,288.39+(-$94.23) = **$10,194.16**
3	1.5	(.02)($10,194.16) = **$203.88**	**$300.00**	$203.88 - $300.00 = **-$96.12**	$10,194.16+(-$96.12) = **$10,098.04**
4	2	(.02)($10,098.04) = **$201.96**	**$300.00**	$201.96 - $300.00 = **-$98.04**	$10,094.04+(-$98.04) = **$10,000**

The Pack realized that construction of these schedules was similar to creating a mortgage payment schedule. "You have a nice check on your work," said Roger, "since the schedules should balance out to the face value of the bond if done correctly." He then displayed two interactive spreadsheets showing complete bond amortization schedules. Here is the first one:

Spreadsheet by ROGER					
Topic: Bond Amortization where COUPON RATE > YIELD RATE					
$50,000	Enter face value of bond				
10	Enter bond term (years)				
6.00%	Enter COUPON rate (compounded semiannually)				
5.00%	Enter YIELD rate (compounded semiannually)				
$1,500.00	Value of semiannual coupon				
$53,897.29	Price paid for bond				
		A	C	A - C	
Coupon #	Year #	Interest required	Coupon	Book value change	Book value
0	0				$53,897.29
1	0.5	$1,347.43	$1,500.00	-$152.57	$53,744.72
2	1	$1,343.62	$1,500.00	-$156.38	$53,588.34
3	1.5	$1,339.71	$1,500.00	-$160.29	$53,428.05
4	2	$1,335.70	$1,500.00	-$164.30	$53,263.75
5	2.5	$1,331.59	$1,500.00	-$168.41	$53,095.34
6	3	$1,327.38	$1,500.00	-$172.62	$52,922.73
7	3.5	$1,323.07	$1,500.00	-$176.93	$52,745.80
8	4	$1,318.64	$1,500.00	-$181.36	$52,564.44
9	4.5	$1,314.11	$1,500.00	-$185.89	$52,378.55
10	5	$1,309.46	$1,500.00	-$190.54	$52,188.02
11	5.5	$1,304.70	$1,500.00	-$195.30	$51,992.72
12	6	$1,299.82	$1,500.00	-$200.18	$51,792.53
13	6.5	$1,294.81	$1,500.00	-$205.19	$51,587.35
14	7	$1,289.68	$1,500.00	-$210.32	$51,377.03
15	7.5	$1,284.43	$1,500.00	-$215.57	$51,161.46
16	8	$1,279.04	$1,500.00	-$220.96	$50,940.49
17	8.5	$1,273.51	$1,500.00	-$226.49	$50,714.01
18	9	$1,267.85	$1,500.00	-$232.15	$50,481.86
19	9.5	$1,262.05	$1,500.00	-$237.95	$50,243.90
20	10	$1,256.10	$1,500.00	-$243.90	$50,000.00

In this situation the desired yield rate is less than the coupon rate. The book value decreases with each coupon payment since the excess interest makes the bond value lessen for the investor. Herkimer did point out that the bond could be sold at any time for whatever price it would bring, but the book value as calculated is important when it comes to listing assets and for tax purposes.

As the final part of his project, Roger displayed the complete amortization schedule for a 20-year bond in which the coupon rate was less than the desired yield rate.

Spreadsheet by ROGER					
Topic: Bond Amortization where COUPON RATE < YIELD RATE					
$100,000	Enter face value of bond				
20	Enter bond term (years)				
5.00%	Enter COUPON rate (compounded semiannually)				
6.50%	Enter YIELD rate (compounded semiannually)				
$2,500.00	Value of semiannual coupon				
$83,343.68	Price paid for bond				

		A	C	A - C	
Coupon #	Year #	Interest required	Coupon	Book value change	Book value
0	0				$83,343.68
1	0.5	$2,708.67	$2,500.00	$208.67	$83,552.34
2	1	$2,715.45	$2,500.00	$215.45	$83,767.80
3	1.5	$2,722.45	$2,500.00	$222.45	$83,990.25
4	2	$2,729.68	$2,500.00	$229.68	$84,219.93
5	2.5	$2,737.15	$2,500.00	$237.15	$84,457.08
6	3	$2,744.86	$2,500.00	$244.86	$84,701.94
7	3.5	$2,752.81	$2,500.00	$252.81	$84,954.75
8	4	$2,761.03	$2,500.00	$261.03	$85,215.78
9	4.5	$2,769.51	$2,500.00	$269.51	$85,485.29
10	5	$2,778.27	$2,500.00	$278.27	$85,763.56
11	5.5	$2,787.32	$2,500.00	$287.32	$86,050.88
12	6	$2,796.65	$2,500.00	$296.65	$86,347.53
13	6.5	$2,806.29	$2,500.00	$306.29	$86,653.83
14	7	$2,816.25	$2,500.00	$316.25	$86,970.08
15	7.5	$2,826.53	$2,500.00	$326.53	$87,296.60
16	8	$2,837.14	$2,500.00	$337.14	$87,633.74
17	8.5	$2,848.10	$2,500.00	$348.10	$87,981.84
18	9	$2,859.41	$2,500.00	$359.41	$88,341.25
19	9.5	$2,871.09	$2,500.00	$371.09	$88,712.34
20	10	$2,883.15	$2,500.00	$383.15	$89,095.49
21	10.5	$2,895.60	$2,500.00	$395.60	$89,491.09
22	11	$2,908.46	$2,500.00	$408.46	$89,899.55
23	11.5	$2,921.74	$2,500.00	$421.74	$90,321.29
24	12	$2,935.44	$2,500.00	$435.44	$90,756.73
25	12.5	$2,949.59	$2,500.00	$449.59	$91,206.33
26	13	$2,964.21	$2,500.00	$464.21	$91,670.53
27	13.5	$2,979.29	$2,500.00	$479.29	$92,149.82
28	14	$2,994.87	$2,500.00	$494.87	$92,644.69
29	14.5	$3,010.95	$2,500.00	$510.95	$93,155.65
30	15	$3,027.56	$2,500.00	$527.56	$93,683.20
31	15.5	$3,044.70	$2,500.00	$544.70	$94,227.91
32	16	$3,062.41	$2,500.00	$562.41	$94,790.31
33	16.5	$3,080.69	$2,500.00	$580.69	$95,371.00
34	17	$3,099.56	$2,500.00	$599.56	$95,970.56
35	17.5	$3,119.04	$2,500.00	$619.04	$96,589.60
36	18	$3,139.16	$2,500.00	$639.16	$97,228.76
37	18.5	$3,159.93	$2,500.00	$659.93	$97,888.70
38	19	$3,181.38	$2,500.00	$681.38	$98,570.08
39	19.5	$3,203.53	$2,500.00	$703.53	$99,273.61
40	20	$3,226.39	$2,500.00	$726.39	$100,000.00

Roger received deserved applause for his presentation.

HERKIMER'S FASCINATING FINANCIAL FACTS:

In 2004 subtle background colors were introduced on bills. For instance, on the $20 bill the background colors are green, blue and peach. A blue eagle is printed in blue on the front along with the blue TWENTY USA words to the right of the portrait of Jackson. The back of the bill contains small yellow numeral 20s in the background. The use of colors make the bills more difficult to counterfeit.

ACTIVITY SET FOR SESSION #39

1. **ALCAL** activity: Use your calculator to check the computations bond purchase prices displayed on the two spreadsheets produced by Roger:

   (a) The $50,000 ten-year bond with 6% semiannual coupons purchased to yield 5% compounded semiannually:

To be evaluated	$1500[1-(1+.05/2)^{-20}]/(.05/2)+50000(1+.05/2)^{-20}$
Calculator entry	1500*(1-(1+.05/2)^~20)/(.05/2)+50000*(1+.05/2)^~20
Number represented	

   (b) The $100,000 twenty-year bond with 5% semiannual coupons purchased to yield 6.5% compounded semiannually:

To be evaluated	$2500[1-(1+.065/2)^{-40}]/(.065/2)+100000(1+.05/2)^{-40}$
Calculator entry	2500*(1-(1+.065/2)^~40)/(.065/2)+100000*(1+.065/2)^~40
Number represented	

2. REFLECTION & COMMENT activity:

   Some people confuse amortized loans with interest-only loans. And interest-only loan is exactly what it sounds like. Simply put, the entire amount of your scheduled payments goes to the interest due on the loan; your scheduled payments do not go towards the principal of the loan. You may, however, make additional specific principal payments. These loans can be beneficial since they generally allow for smaller payments. With an amortized loan, it is true that the largest part of the payment goes towards interest - at least in the beginning of the loan - but a portion of the principal is in fact, paid down as well.

   (Sherry Holzhy, *What is Amortization?* Wisegeek.com)

3. If a 25-year $500,000 bond with 7% semiannual coupons was purchased to yield 7%, explain why the book value after any coupon payment would be $500,000.

4. Complete the amortization schedule for a 3-year $200,000 bond with 8% semiannual coupons purchased to yield an annual rate of 10% compounded semiannually:

Coupon #	Year #	A = Interest Required	C = Coupon	Book Value Change (A-C)	Book Value
0	0				
1	0.5				
2	1				
3	1.5				
4	2				
5	2.5				
6	3				

5. Complete the amortization schedule for a 3-year $200,000 bond with 8% semiannual coupons purchased to yield an annual rate of 5% compounded semiannually:

Coupon #	Year #	A = Interest Required	C = Coupon	Book Value Change (A-C)	Book Value
0	0				
1	0.5				
2	1				
3	1.5				
4	2				
5	2.5				
6	3				

$$
$$$

Session #40
STEPHEN'S PROJECT: INFLATION CONCERNS IN RETIREMENT

> *Honesty is the recognition of the fact that neither love nor fame nor cash is a value if obtained by fraud.*
> -Ayn Rand

    During informal discussions with the Pack, Herkimer frequently stressed that with any long-term investment *inflation* is always a lurking factor. The group knew from previous sessions that inflation erodes the purchasing power of money. Stephen had the challenge to demonstrate why inflation should be a definite consideration in retirement planning. He started by putting this spreadsheet on the overhead projector. He explained that the first chart displayed the accumulation of a single payment of $1000 for 10 years at a true rate of 4%. The second chart illustrated the effects of an annual inflation rate of 4%.

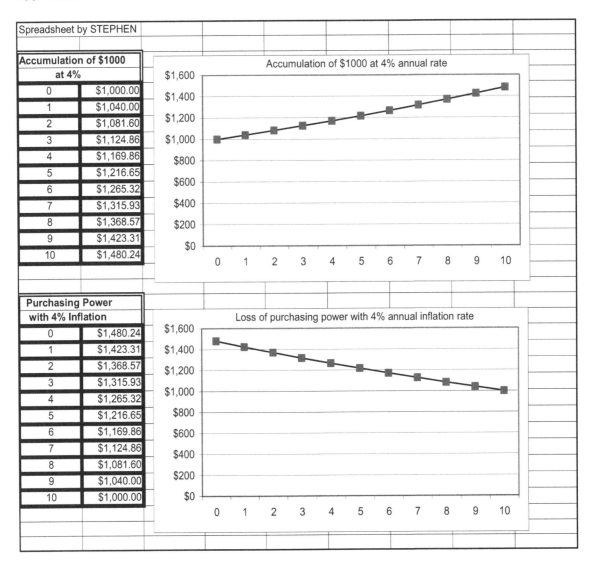

Accumulation of $1000 at 4%	
0	$1,000.00
1	$1,040.00
2	$1,081.60
3	$1,124.86
4	$1,169.86
5	$1,216.65
6	$1,265.32
7	$1,315.93
8	$1,368.57
9	$1,423.31
10	$1,480.24

Purchasing Power with 4% Inflation	
0	$1,480.24
1	$1,423.31
2	$1,368.57
3	$1,315.93
4	$1,265.32
5	$1,216.65
6	$1,169.86
7	$1,124.86
8	$1,081.60
9	$1,040.00
10	$1,000.00

Stephen assisted the Pack with a bit of review. He pointed out that if $1000 accumulates at a true annual rate of 4%, then the accumulation values are simply $1000, $1000(1.04), $1000(1.04)^2, ..., $1000(1.04)^{10}$. The last value is $1480.24, as indicated in the spreadsheet table.

"Now," said Stephen, "let's assume that we have $1480.24 in an envelope stuffed away somewhere in our house for future use. If inflation is 4% a year and we simply let this money remain unspent, then in 10 years it will buy only $1000 worth of presently-priced items." He went on to explain that the $1480.24 shrunk to $1000 after 10 years at a 4% inflation rate. The tabular values in the spreadsheet are simply $1480.24, $1480.24/(1.04)$, $1480.24/(1.04)^2, ..., $1480.24/(1.04)^{10}$. The last value is $1000.

Stephen went on to explain that he learned that some people do not take inflation into account when they think of retirement. "If this is not done," he said, "retirement plans might not work out as one might hope." He asked the Pack to now consider this scenario: Suppose you have $500,000 in a retirement fund that is earning a true annual rate of 5% and you want to withdraw $35,000 a year until the fund is depleted. He said he would show that under these circumstances you could withdraw this amount for 28 years. But with inflation, the $35,000 you receive 28 years from now will not buy $35,000 worth of present day items. If, for instance, inflation is 3% a year then the last complete payment withdrawn will buy only $35,000/(1.03)^{28} = $15,298$ worth of present day items.

Stephen went on to explain that you might want to have your annual withdrawals increase a bit each year to compensate for inflation. A plan might be to start off with a first-year withdrawal of $35,000 and then have future annual withdrawals increase by 5%. With this plan your first withdrawal would be $35,000, the second would be $35,000(1.05) = $36,750, the third would be $35,000(1.05)^2 = $38,588$, etc. This withdrawal plan would have your funds lasting just 15 years.

To establish the validity of what he had stated Stephen put this spreadsheet on the overhead:

Spreadsheet by STEPHEN								

$500,000 <---Initial fund amount

6% <--True annual interest rate earned by fund

$35,000 <---Amount withdrawn each year   **5%** <--- Inflation factor applied to annual withdrawals

Year	Amount in fund	Amount withdrawn	Fund amount remaining	Fund amount with accumulated interest	Year	Amount in fund	Amount withdrawn	Fund amount remaining	Fund amount with accumulated interest
1	$500,000	$35,000	$465,000	$492,900	1	$500,000	$35,000	$465,000	$492,900
2	$492,900	$35,000	$457,900	$485,374	2	$492,900	$36,750	$456,150	$483,519
3	$485,374	$35,000	$450,374	$477,396	3	$483,519	$38,588	$444,932	$471,627
4	$477,396	$35,000	$442,396	$468,940	4	$471,627	$40,517	$431,111	$456,977
5	$468,940	$35,000	$433,940	$459,977	5	$456,977	$42,543	$414,434	$439,300
6	$459,977	$35,000	$424,977	$450,475	6	$439,300	$44,670	$394,631	$418,308
7	$450,475	$35,000	$415,475	$440,404	7	$418,308	$46,903	$371,405	$393,689
8	$440,404	$35,000	$405,404	$429,728	8	$393,689	$49,249	$344,441	$365,107
9	$429,728	$35,000	$394,728	$418,412	9	$365,107	$51,711	$313,396	$332,200
10	$418,412	$35,000	$383,412	$406,416	10	$332,200	$54,296	$277,904	$294,578
11	$406,416	$35,000	$371,416	$393,701	11	$294,578	$57,011	$237,567	$251,821
12	$393,701	$35,000	$358,701	$380,223	12	$251,821	$59,862	$191,959	$203,476
13	$380,223	$35,000	$345,223	$365,937	13	$203,476	$62,855	$140,621	$149,059
14	$365,937	$35,000	$330,937	$350,793	14	$149,059	$65,998	$83,061	$88,045
15	$350,793	$35,000	$315,793	$334,741	15	$88,045	$69,298	$18,747	$19,872
16	$334,741	$35,000	$299,741	$317,725	16	$19,872	$72,762	($52,891)	**($56,064)**
17	$317,725	$35,000	$282,725	$299,689	17	($56,064)	$76,401	($132,465)	**($140,413)**
18	$299,689	$35,000	$264,689	$280,570	18	($140,413)	$80,221	($220,633)	**($233,871)**
19	$280,570	$35,000	$245,570	$260,304	19	($233,871)	$84,232	($318,103)	**($337,189)**
20	$260,304	$35,000	$225,304	$238,822	20	($337,189)	$88,443	($425,632)	**($451,170)**
21	$238,822	$35,000	$203,822	$216,052	21	($451,170)	$92,865	($544,036)	**($576,678)**
22	$216,052	$35,000	$181,052	$191,915	22	($576,678)	$97,509	($674,187)	**($714,638)**
23	$191,915	$35,000	$156,915	$166,330	23	($714,638)	$102,384	($817,022)	**($866,043)**
24	$166,330	$35,000	$131,330	$139,209	24	($866,043)	$107,503	($973,547)	**($1,031,959)**
25	$139,209	$35,000	$104,209	$110,462	25	($1,031,959)	$112,878	($1,144,838)	**($1,213,528)**
26	$110,462	$35,000	$75,462	$79,990	26	($1,213,528)	$118,522	($1,332,051)	**($1,411,974)**
27	$79,990	$35,000	$44,990	$47,689	27	($1,411,974)	$124,449	($1,536,422)	**($1,628,608)**
28	$47,689	$35,000	$12,689	$13,450	28	($1,628,608)	$130,671	($1,759,278)	**($1,864,835)**
29	$13,450	$35,000	($21,550)	**($22,843)**	29	($1,864,835)	$137,205	($2,002,040)	**($2,122,162)**
30	($22,843)	$35,000	($57,843)	**($61,313)**	30	($2,122,162)	$144,065	($2,266,227)	**($2,402,200)**

The Pack was quick to note that Stephen's statements were correct. With inflation not taken into consideration after 28 years there would be only $13,450 left, not enough for a complete payment of $35,000 in year 29. With payments increasing 5% a year to compensate for inflation, there would be $19,872 remaining after 16 years.

Stephen displayed another example for Herkimer and the Pack. In this case the initial fund was $1,000,000 and earned a true annual rate of 5%. The initial annual withdrawal was $60,000. The table on the right displayed the situation when a 3% inflation rate was introduced and payments were increased by 3% to compensate for the potential loss of purchasing power.

Spreadsheet by STEPHEN

$1,000,000 <---Initial fund amount

5% <--True annual interest rate earned by fund

$60,000 <---Amount withdrawn each year   3% <--- Inflation factor applied to annual withdrawals

Year	Amount in fund	Amount withdrawn	Fund amount remaining	Fund amount with accumulated interest	Year	Amount in fund	Amount withdrawn	Fund amount remaining	Fund amount with accumulated interest
1	$1,000,000	$60,000	$940,000	$987,000	1	$1,000,000	$60,000	$940,000	$987,000
2	$987,000	$60,000	$927,000	$973,350	2	$987,000	$61,800	$925,200	$971,460
3	$973,350	$60,000	$913,350	$959,018	3	$971,460	$63,654	$907,806	$953,196
4	$959,018	$60,000	$899,018	$943,968	4	$953,196	$65,564	$887,633	$932,014
5	$943,968	$60,000	$883,968	$928,167	5	$932,014	$67,531	$864,484	$907,708
6	$928,167	$60,000	$868,167	$911,575	6	$907,708	$69,556	$838,152	$880,059
7	$911,575	$60,000	$851,575	$894,154	7	$880,059	$71,643	$808,416	$848,837
8	$894,154	$60,000	$834,154	$875,862	8	$848,837	$73,792	$775,044	$813,797
9	$875,862	$60,000	$815,862	$856,655	9	$813,797	$76,006	$737,790	$774,680
10	$856,655	$60,000	$796,655	$836,487	10	$774,680	$78,286	$696,393	$731,213
11	$836,487	$60,000	$776,487	$815,312	11	$731,213	$80,635	$650,578	$683,107
12	$815,312	$60,000	$755,312	$793,077	12	$683,107	$83,054	$600,053	$630,056
13	$793,077	$60,000	$733,077	$769,731	13	$630,056	$85,546	$544,510	$571,736
14	$769,731	$60,000	$709,731	$745,218	14	$571,736	$88,112	$483,624	$507,805
15	$745,218	$60,000	$685,218	$719,479	15	$507,805	$90,755	$417,049	$437,902
16	$719,479	$60,000	$659,479	$692,453	16	$437,902	$93,478	$344,424	$361,645
17	$692,453	$60,000	$632,453	$664,075	17	$361,645	$96,282	$265,363	$278,631
18	$664,075	$60,000	$604,075	$634,279	18	$278,631	$99,171	$179,460	$188,433
19	$634,279	$60,000	$574,279	$602,993	19	$188,433	$102,146	$86,287	$90,601
20	$602,993	$60,000	$542,993	$570,143	20	$90,601	$105,210	($14,609)	($15,340)
21	$570,143	$60,000	$510,143	$535,650	21	($15,340)	$108,367	($123,706)	($129,892)
22	$535,650	$60,000	$475,650	$499,432	22	($129,892)	$111,618	($241,509)	($253,585)
23	$499,432	$60,000	$439,432	$461,404	23	($253,585)	$114,966	($368,551)	($386,979)
24	$461,404	$60,000	$401,404	$421,474	24	($386,979)	$118,415	($505,394)	($530,663)
25	$421,474	$60,000	$361,474	$379,548	25	($530,663)	$121,968	($652,631)	($685,263)
26	$379,548	$60,000	$319,548	$335,525	26	($685,263)	$125,627	($810,889)	($851,434)
27	$335,525	$60,000	$275,525	$289,301	27	($851,434)	$129,395	($980,829)	($1,029,871)
28	$289,301	$60,000	$229,301	$240,766	28	($1,029,871)	$133,277	($1,163,148)	($1,221,305)
29	$240,766	$60,000	$180,766	$189,805	29	($1,221,305)	$137,276	($1,358,581)	($1,426,510)
30	$189,805	$60,000	$129,805	$136,295	30	($1,426,510)	$141,394	($1,567,904)	($1,646,299)
31	$136,295	$60,000	$76,295	$80,110	31	($1,646,299)	$145,636	($1,791,935)	($1,881,532)
32	$80,110	$60,000	$20,110	$21,115	32	($1,881,532)	$150,005	($2,031,537)	($2,133,113)
33	$21,115	$60,000	($38,885)	($40,829)	33	($2,133,113)	$154,505	($2,287,618)	($2,401,999)
34	($40,829)	$60,000	($100,829)	($105,870)	34	($2,401,999)	$159,140	($2,561,139)	($2,689,196)
35	($105,870)	$60,000	($165,870)	($174,164)	35	($2,689,196)	$163,914	($2,853,111)	($2,995,766)
36	($174,164)	$60,000	($234,164)	($245,872)	36	($2,995,766)	$168,832	($3,164,598)	($3,322,828)

As the spreadsheet indicates, the constant payments of $60,000 will last 32 years. With an inflation rate of 3% taken into consideration, the fund can support the increasing annual payments for 19 years.

"Great job, Stephen" concluded Herkimer. "Retirement income is important, and you have clearly indicated and illustrated factors that must be considered in planning for one's future welfare."

219

ACTIVITY SET FOR VIGNETTE #40

1. **ALCAL** activity: Use your calculator to check the following computations on the spreadsheets produced by Stephen:

(a) The amount withdrawn in 11 years on a retirement fund to compensate for a 5% inflation rate when the initial amount withdrawn is $35,000:

To be evaluated	Calculator entry	Number represented
$35000(1.05)^{10}$	35000*1.05^10	

(b) The amount withdrawn in 16 years on a retirement fund to compensate for a 3% inflation rate when the initial amount withdrawn is $60,000:

To be evaluated	Calculator entry	Number represented
$60000(1.03)^{15}$	60000*1.03^15	

2. REFLECTION & COMMENT activity:

Most people underestimate the impact inflation will have on their retirement plans. Even at relatively low rates, inflation is a real thief of buying power over time. Most experts feel safe recommending that individuals calculate their retirement needs using a 3 percent inflation rate. But it is important to understand that we have seen (as in the late seventies and early eighties) sustained inflation rates of around 10 percent.

(NewRetirement.com, June 2009)

3. Research the topic of *retirement saving* on the internet.

4. Have a group discussion about the following statement in an article produced by Charles Schwab and Co., Inc.

> If you start saving roughly 12% of your income in your mid-20s and have the discipline to maintain your savings plan, you shouldn't have to increase that percentage as you go through your 30s, 40s and so on.
>
> That's the benefit of getting an early start saving for retirement: The lower-percentage savings guideline stays with you the rest of your working life, if you can stick with it. The later you get started, the higher the percentage of your income you'll need to save.
>
> Of course, the paradox is that during our 20s, when we can lock in the lowest savings rate, we're not thinking about retirement. We're paying off student loans, saving for our first home, buying furniture and home entertainment systems, and generally having fun. Then, by the time we get past that stage and start getting serious about the future, we've grown so comfortable in our lifestyles that the percentage of income we need to save seems beyond our reach.

5. Consider a $600,000 retirement fund that will earn a true annual rate of 5.5%. Payments of $40,000 will be withdrawn from this fund on an annual basis. How long will the fund last if (a) the payments remain constant at $40,000? (b) the payments are increased by 2% annually to compensate for inflation? (c) the payments are increased by 4% annually to compensate for inflation?

6. An initial payment of $50,000 will be withdrawn from a $900,000 retirement fund. Future annual payments will be increased by 3.5% to compensate for anticipated inflation. How long will the fund last if it earns a true annual rate of (a) 4%? (b) 6%? (c) 8%?

$$
$$

Session #41
GLEN'S PROJECT: TEASER RATES

> *A low teaser rate can buy you time to pay off your balance - but read the fine print.*
> -Thomas Anderson

The concept of a *teaser rate* was not familiar to Glen but he understood it almost immediately when starting his project since credit card debt had been a previous discussion topic for the Pack. "A *teaser rate*," he explained, "is a very low initial rate, perhaps even 0%, that lasts for a very limited time. After that, the interest rate automatically goes up, sometimes drastically." He went on to say a credit card company might try to get you to transfer your debt from another card to their card by offering a low teaser rate.

Glen demonstrated how a teaser rate works by asking the group to consider the following situation: Suppose you have $2,000 in credit card debt and transfer it to a card that has a 0% rate for six months. After the six months, the rate increases to an annual rate of 24%, compounded monthly. He put the following transparency on the overhead and asked the Pack to note what begins to happen with the 7th payment if one makes just a minimum payment of $40 a month:

MONTH	PAYMENT	INTEREST	PRINCIPAL	STILL OWE
0	/////	/////	/////	$2000.00
1	$40	$0.00	$40.00	$1960.00
2	$40	$0.00	$40.00	$1920.00
3	$40	$0.00	$40.00	$1880.00
4	$40	$0.00	$40.00	$1840.00
5	$40	$0.00	$40.00	$1800.00
6	$40	$0.00	$40.00	$1760.00
7	$40	$1760(.02) = $35.20	$4.80	$1755.20
8	$40	$1755.20(.02) = $35.10	$4.90	$1750.30

The Pack observed that the interest would begin "piling up" beginning with the 7th payment. Glen then noted that a teaser rate isn't necessarily a bad deal. "Suppose," he said, "you could transfer the $2,000 from a card with a high interest rate to one with a 0% teaser rate for 6 months. If you could pay off that balance within 6 months, say with 6 monthly payments of about $335 a month, this could be a good deal and save you a lot of

222

money. As Herkimer has indicated to us many times, it pays to be savvy about your personal financial situation."

Glen then put up a spreadsheet he had constructed that displayed two years of payments if one simply paid a minimum of $40 a month with the teaser rate of 0%:

Spreadsheet by GLEN						
Topic: TEASER RATES (for 6 months)						
Rates are annual rates compounded monthly						
$2,000	<--- Amount of debt					
0%	<---TEASER RATE					
24%	<---Rate after 6 months of TEASER RATE					
$40	<---Monthly payment made					
Month #	Payment	Interest	Principal	Still Owe		
0				$2,000.00		
1	$40	$0.00	$40.00	$1,960.00		
2	$40	$0.00	$40.00	$1,920.00		
3	$40	$0.00	$40.00	$1,880.00		
4	$40	$0.00	$40.00	$1,840.00		
5	$40	$0.00	$40.00	$1,800.00		
6	$40	$0.00	$40.00	$1,760.00		
7	$40	$35.20	$4.80	$1,755.20	<--End of TEASER	
8	$40	$35.10	$4.90	$1,750.30		
9	$40	$35.01	$4.99	$1,745.31		
10	$40	$34.91	$5.09	$1,740.22		
11	$40	$34.80	$5.20	$1,735.02		
12	$40	$34.70	$5.30	$1,729.72		
13	$40	$34.59	$5.41	$1,724.32		
14	$40	$34.49	$5.51	$1,718.80		
15	$40	$34.38	$5.62	$1,713.18		
16	$40	$34.26	$5.74	$1,707.44		
17	$40	$34.15	$5.85	$1,701.59		
18	$40	$34.03	$5.97	$1,695.62		
19	$40	$33.91	$6.09	$1,689.53		
20	$40	$33.79	$6.21	$1,683.33		
21	$40	$33.67	$6.33	$1,676.99		
22	$40	$33.54	$6.46	$1,670.53		
23	$40	$33.41	$6.59	$1,663.94		
24	$40	$33.28	$6.72	$1,657.22		
	$960	$617.22	$342.78		<-Two-year totals	

This startled the Pack. After 2 years with the 6-month 0% teaser rate, a total of $960 had been paid and only $342.78 of the loan had been paid off. The outstanding balance was $1,657.22.

Glen produced another spreadsheet example. In this scenario, the teaser rate was 2%. This would increase to 30% after one year. The rates are annual rates, compounded monthly. The scheduled outlined 3 years assuming a monthly payment of $200.

Spreadsheet by GLEN						
Topic: TEASER RATES (for 12 months)						
Rates are annual rates compounded monthly						
$10,000	<--- Amount of debt					
2%	<---TEASER RATE					
30%	<---Rate after 1 year of TEASER RATE					
$200	<---Monthly payment made					
Month #	Payment	Interest	Principal	Still Owe		
0				$10,000.00		
1	$200	$16.67	$183.33	$9,816.67		
2	$200	$16.36	$183.64	$9,633.03		
3	$200	$16.06	$183.94	$9,449.08		
4	$200	$15.75	$184.25	$9,264.83		
5	$200	$15.44	$184.56	$9,080.27		
6	$200	$15.13	$184.87	$8,895.41		
7	$200	$14.83	$185.17	$8,710.23		
8	$200	$14.52	$185.48	$8,524.75		
9	$200	$14.21	$185.79	$8,338.96		
10	$200	$13.90	$186.10	$8,152.86		
11	$200	$13.59	$186.41	$7,966.44		
12	$200	$13.28	$186.72	$7,779.72		
13	$200	$194.49	$5.51	$7,774.21	<--- End TEASER RATE	
14	$200	$194.36	$5.64	$7,768.57		
15	$200	$194.21	$5.79	$7,762.78		
16	$200	$194.07	$5.93	$7,756.85		
17	$200	$193.92	$6.08	$7,750.77		
18	$200	$193.77	$6.23	$7,744.54		
19	$200	$193.61	$6.39	$7,738.16		
20	$200	$193.45	$6.55	$7,731.61		
21	$200	$193.29	$6.71	$7,724.90		
22	$200	$193.12	$6.88	$7,718.02		
23	$200	$192.95	$7.05	$7,710.97		
24	$200	$192.77	$7.23	$7,703.75		
25	$200	$192.59	$7.41	$7,696.34		
26	$200	$192.41	$7.59	$7,688.75		
27	$200	$192.22	$7.78	$7,680.97		
28	$200	$192.02	$7.98	$7,672.99		
29	$200	$191.82	$8.18	$7,664.82		
30	$200	$191.62	$8.38	$7,656.44		
31	$200	$191.41	$8.59	$7,647.85		
32	$200	$191.20	$8.80	$7,639.05		
33	$200	$190.98	$9.02	$7,630.02		
34	$200	$190.75	$9.25	$7,620.77		
35	$200	$190.52	$9.48	$7,611.29		
36	$200	$190.28	$9.72	$7,601.58		
	$7,200	$4,802	$2,398		<-Three-year totals	

"If you thought you could pay off most or all of the $10,000 within a year, the teaser rate of 2% might be a good deal," said Glen. ""But the change at 13 months is drastic if you can pay only $200 a month. After three years you have paid over $4,800 in interest and you still owe over $7,600."

"Glen has clearly demonstrated that teaser rates can lead one right into the credit card trap we previously discussed," concluded Herkimer.

ACTIVITY SET FOR SESSION #41

1. **ALCAL** activity: Use your calculator to check the following computations on the spreadsheets produced by Glen:

(a) On the first spreadsheet the amount of interest in the payment of $40 at month #7:

To be evaluated	Calculator entry	Number represented
1760(.24/12)	1760*.24/12	

(b) On the second spreadsheet the amount of interest in the payment of $200 at month # 13:

To be evaluated	Calculator entry	Number represented
7779.72(.30/12)	7779.72*.30/12	

2. REFLECTION & COMMENT activity:

A teaser rate is an artificially low initial rate which lasts only for a limited time and often for limited charges, such as transfers of balances from other cards. Most teaser rates are good only for six months or less. After that, the rate automatically goes up. Remember that if you build up a balance and pay it after the period of a temporary rate, the much higher permanent rate will apply to your payment plan. This means that the permanent long term rate on the card is much more important than the temporary rate.

(Bond & Botes PC, Life *After Bankruptcy*, 2007))

3. Use the internet to research the concept of a teaser rate on credit cards. Doing so should convince you that you should definitely "read the fine print" before agreeing to any debt transfer to take advantage of a teaser rate.

4. Find advertisements in newspapers, magazines, and on television where teaser rates are being offered. They are very common.

5. Consider $8,000 in credit card debt that is transferred to a card offering a 1% teaser rate for 3 months before a 22% annual rate, compounded monthly, begins to be applied to the outstanding balance. If a payment of $150 a month is made, complete the following table to display for the first six months of this financial plan:

Month	Payment	Interest	Principal	Still Owe
0				$8,000.00
1	$150.00			
2	$150.00			
3	$150.00			
4	$150.00			
5	$150.00			
6	$150.00			

6. Consider $12,000 in credit card debt that is transferred to a card offering a 0% teaser rate for 6 months. After this period, an annual rate of 28%, compounded monthly, is to be applied to the outstanding balance. Payments of $300 a month will be made. Hence, after 6 months the outstanding balance would be $12,000 - 6($300) = $10,200. For the 7th payment calculate the interest and principal portions and the outstanding balance after the payment. Do the same for the 8th payment.

$$
$$

Session #42
JANICE'S PROJECT: ADJUSTABLE RATE MORTGAGES (ARMs)

> *October: This is one of the most dangerous months to invest in stocks. Other dangerous months are July, January, September, April, November, May, March, June, December, August and February.*
> -Mark Twain

Janice began by noting that the previous mortgages examined and created by the Pack were *Fixed Rate Mortgages* (FRMs). The loan rate was constant throughout the term of the mortgage. With an *Adjustable Rate Mortgage* (ARM) the interest rate is adjusted during the loan term after specific amounts of time, often a period of 1 year or a period of 5 years. This means that the monthly mortgage payment can vary from year to year.

"I learned that ARMs are attractive because the initial rate is often significantly lower than that of a FRM," said Janice. "Banks can offer lower rates on ARMs because it doesn't lock them into a constant rate for a long period of time. However you really have to do some research before signing on to an ARM because the monthly payments can increase by quite a bit when the time for the initial rate expires." She went on to explain that changes in the rate are generally a function of an "index" that should be disclosed at the inception of the ARM. A commonly used index is the Federal Reserve Funds Index. "I don't want to get bogged down with the index concept," she continued, "but it is really important for any borrower to understand what can trigger a change in a loan interest rate."

"And banks are businesses," continued Janice. "The concept of a *teaser rate* applies to ARMs as well as credit cards. A bank might entice you with a very low initial mortgage rate on an ARM. A lot of people get carried away with the thought of a mortgage with a low monthly payment and fail to think about what happens when the initial rate increases as it almost always does. As Herkimer has frequently told us, it's important to really understand what is going on with your money."

Janice stated that ARMs have various forms. She put this transparency on the overhead:

Some ARM Types	
1/1 ARM	*Initial rate set for 1 year; thereafter can be adjusted after 1 year*
5/1 ARM	*Initial rate set for 5 years; thereafter can be adjusted after 1 year*
5/5 ARM	*Initial rate set for 5 years; thereafter can be adjusted every 5 years*

After a brief period of time she then put up a transparency containing some hand-written computation she had done relating to a 30-year $400,000 ARM (1/1) with an initial rate of 6%. The transparency displayed a schedule where the rate increased to 7% for the second year and 8% for the third year.

$400,000 ARM (1/1), 30 YEARS, INITIAL RATE = 6%.

MONTHLY PYMT. $= \dfrac{\$400,000\left(.06/12\right)}{1-\left(1+\frac{.06}{12}\right)^{-360}} = \$2398.20.$

PRINCIPAL OUTSTANDING AFTER 12 MONTHS

$$= \$2398.20\left[\dfrac{1-\left(1+.06/12\right)^{-348}}{.06/12}\right]$$

$$= \$395,087.61.$$

1% INCREASE

THIS IS NOW A "NEW" 29-YEAR LOAN AT 7%.

MONTHLY PYMT. $= \dfrac{\left(\$395,087.61\right)\left(.07/12\right)}{1-\left(1+\frac{.07}{12}\right)^{-348}} = \$2655.50.$

PRINCIPAL OUTSTANDING AFTER 24 MONTHS

$$= \$2655.50\left[\dfrac{1-\left(1+.07/12\right)^{-336}}{.07/12}\right]$$

$$= \$390,739.60.$$

1% INCREASE

THIS IS NOW A "NEW" 28-YEAR LOAN AT 8%.

MONTHLY PYMT. $= \dfrac{\$390,739.60\left(.08/12\right)}{1-\left(1+\frac{.08}{12}\right)^{-336}}$

$$= \$2917.88$$

SUMMARY: TO NEAREST DOLLAR, 1ST YR. PYMT. = $2398,
2nd YR. PYMT. = $2656,
3rd YR. PYMT. = $2918.

Janice then put up a spreadsheet displaying the first three years of this 1/1 ARM schedule. "You can see why you would want to really understand any ARM agreement you sign," she said. "Perhaps you can afford $2398 during the first year, but will you be able to afford an additional $500 a month in the third year?"

Spreadsheet by JANICE					
ARMs (Adjustable Rate Mortgages)					
Example of a 1/1 ARM (Initial rate set for 1 year, can be adjusted every year thereafter)					
$400,000	<---Amount of mortgage loan				
30	<---Mortgage term (years)				
6%	<-- Initial rate				
7%	<---Rate after 1 year				
8%	<---Rate after 2 years				
$2,398.20	<-- Initial monthly payment				
$2,655.50	<---Monthly payment start of year 2				
$2,917.89	<---Monthly payment start of year 3				

Month #	Payment	Interest	Principal	Principal Outstanding	
0				$400,000.00	
1	$2,398.20	$2,000.00	$398.20	$399,601.80	
2	$2,398.20	$1,998.01	$400.19	$399,201.60	
3	$2,398.20	$1,996.01	$402.19	$398,799.41	
4	$2,398.20	$1,994.00	$404.21	$398,395.21	
5	$2,398.20	$1,991.98	$406.23	$397,988.98	
6	$2,398.20	$1,989.94	$408.26	$397,580.72	
7	$2,398.20	$1,987.90	$410.30	$397,170.42	
8	$2,398.20	$1,985.85	$412.35	$396,758.07	
9	$2,398.20	$1,983.79	$414.41	$396,343.66	
10	$2,398.20	$1,981.72	$416.48	$395,927.18	
11	$2,398.20	$1,979.64	$418.57	$395,508.61	
12	$2,398.20	$1,977.54	$420.66	$395,087.95	
13	$2,655.50	$2,304.68	$350.83	$394,737.13	
14	$2,655.50	$2,302.63	$352.87	$394,384.26	
15	$2,655.50	$2,300.57	$354.93	$394,029.33	
16	$2,655.50	$2,298.50	$357.00	$393,672.33	
17	$2,655.50	$2,296.42	$359.08	$393,313.24	
18	$2,655.50	$2,294.33	$361.18	$392,952.07	
19	$2,655.50	$2,292.22	$363.28	$392,588.78	
20	$2,655.50	$2,290.10	$365.40	$392,223.38	
21	$2,655.50	$2,287.97	$367.54	$391,855.84	
22	$2,655.50	$2,285.83	$369.68	$391,486.16	
23	$2,655.50	$2,283.67	$371.84	$391,114.33	
24	$2,655.50	$2,281.50	$374.00	$390,740.32	
25	$2,917.89	$2,604.94	$312.95	$390,427.37	
26	$2,917.89	$2,602.85	$315.04	$390,112.33	
27	$2,917.89	$2,600.75	$317.14	$389,795.19	
28	$2,917.89	$2,598.63	$319.25	$389,475.94	
29	$2,917.89	$2,596.51	$321.38	$389,154.56	
30	$2,917.89	$2,594.36	$323.52	$388,831.04	
31	$2,917.89	$2,592.21	$325.68	$388,505.36	
32	$2,917.89	$2,590.04	$327.85	$388,177.51	
33	$2,917.89	$2,587.85	$330.04	$387,847.47	
34	$2,917.89	$2,585.65	$332.24	$387,515.23	
35	$2,917.89	$2,583.43	$334.45	$387,180.78	
36	$2,917.89	$2,581.21	$336.68	$386,844.10	

Janice then put up a second spreadsheet illustrating a 5/5 ARM. In this situation the mortgage had a rate of 5% that could be adjusted every 5 years. The rate was changed to 7.5% after the first 5 years. Janice wanted to illustrate to the Pack the use of formulas that they had previously developed:

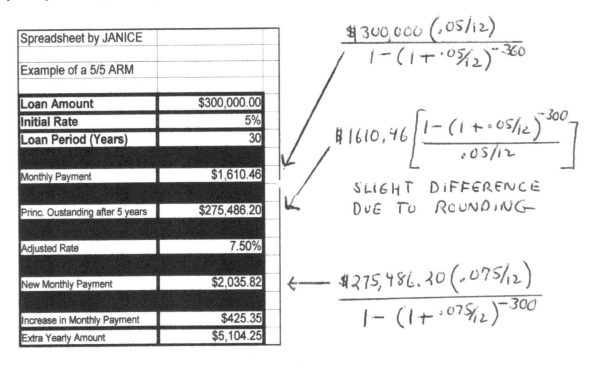

Spreadsheet by JANICE	
Example of a 5/5 ARM	
Loan Amount	$300,000.00
Initial Rate	5%
Loan Period (Years)	30
Monthly Payment	$1,610.46
Princ. Oustanding after 5 years	$275,486.20
Adjusted Rate	7.50%
New Monthly Payment	$2,035.82
Increase in Monthly Payment	$425.35
Extra Yearly Amount	$5,104.25

$$\frac{\$300,000\left(.05/12\right)}{1-\left(1+.05/12\right)^{-360}}$$

$$\$1610.46\left[\frac{1-\left(1+.05/12\right)^{-300}}{.05/12}\right]$$

SLIGHT DIFFERENCE DUE TO ROUNDING

$$\frac{\$275,486.20\left(.075/12\right)}{1-\left(1+.075/12\right)^{-300}}$$

With this 5-year adjustment, the ARM now requires an additional $5,000 a year.

"It is possible that an interest rate could be adjusted downward," said Janice. "Whether a rate is adjusted or not really depends on the financial conditions nationwide at the time of adjustment. It is really, really important for anyone considering an ARM to understand what will produce a rate change, particularly if the initial rate is a teaser rate."

"Nifty job, Janice," said Herkimer. "You've clearly demonstrated that one must fully understand the possible implications resulting from an ARM. Financial literacy is essential if one wants to avoid serious future financial problems."

HERKIMER'S FASCINATING FINANCIAL FACTS:

In 1998 the number of coins produced, by denomination, was as follows:

Pennies	10,257,400,000
Nickels	1,323,672,000
Dimes	2,335,300,000
Quarters	1,867,400,000
Half-dollars	30,710,000

ACTIVITY SET FOR SESSION #42

1. **ALCAL** activity: Use your calculator to check Janice's handwritten computations relating to a 30-year $400,000 ARM (1/1) with an initial rate of 6%:

(a) The initial monthly payment:

To be evaluated	$400000(.06/12)/[1-(1+.06/12)^{-360}]$
Calculator entry	400000*.06/12/(1-(1+.06/12)^360)
Number represented	

(b) The monthly payment when the interest rate became 8% after 2 years:

To be evaluated	$390739.60(.08/12)/[1-(1+.08/12)^{-336}]$
Calculator entry	390739.60*.08/12/(1-(1+.08/12)^336)
Number represented	

2. REFLECTION & COMMENT activity:

An ARM may be an excellent choice if low payments in the near term are your primary requirement or if you don't plan to live in the property long enough for the rates to rise. If interest rates are high and expected to fall, an ARM will ensure that you enjoy lower interest rates without the need to finance. If interest rates are climbing or a steady, predictable payment is important to you, a fix-ed income mortgage may be the way to go.

(James E. McWhinney, *Mortgages: Fixed Rate Versus Adjustable Rate*)

3. Research the topic *adjustable rate mortgages* on the internet. You should be able to find lots of references to the advantages and disadvantages of this type of loan.

IN ACTIVITIES 4-7 COMPLETE THE TABLE FOR EACH INDICATED ARM.

4.

Mortgage Amount	ARM Type	Mortgage Period (Years)	Initial Mortgage Rate	Initial Monthly Payment	Principal Outstanding After 1 Year	Adjusted Monthly Rate	New Monthly Payment
$240,000	1/1	30	5.2%			6.8%	

5.

Mortgage Amount	ARM Type	Mortgage Period (Years)	Initial Mortgage Rate	Initial Monthly Payment	Principal Outstanding After 5 Years	Adjusted Monthly Rate	New Monthly Payment
$320,000	5/1	30	6%			7.6%	

6.

Mortgage Amount	ARM Type	Mortgage Period (Years)	Initial Mortgage Rate	Initial Monthly Payment	Principal Outstanding After 5 Years	Adjusted Monthly Rate	New Monthly Payment
$190,000	5/5	20	5.3%			6.75%	

7.

Mortgage Amount	ARM Type	Mortgage Period (Years)	Initial Mortgage Rate	Initial Monthly Payment	Principal Outstanding After 1 Year	Adjusted Monthly Rate	New Monthly Payment
$420,000	1/1	15	4.9%			6.2%	

$$
$$$

Session #43
THE PACK MAKES A LIST OF USEFUL FINANCIAL FORMULAS

> *My formula for success is to rise early, work late, and strike oil.*
> -J. Paul Getty

Herkimer realized that his time with the Pack was nearing its end. During his numerous sessions with them they had been introduced to a series of financial formulas that allowed them to construct sophisticated financial spreadsheets with relative ease. He now requested that the students "pull them together" in a series of handwritten sheets that they could keep for future reference. "It isn't important that you memorize the formulas," he told them, "but you should know they exist and know where to look for them if you need them in the future."

He reminded the group that if one is going to accumulate or discount payments, then the payment and interest periods must "jive." Beginning in Session #18 the group started working with both true annual rates (annual yields) and nominal annual rates and learned that it was sometimes essential to produce an annual yield equivalent to a specified nominal rate, or to calculate a nominal rate equivalent to a specified annual yield.

Herkimer requested that the Pack start their series of formula sheets with the all-important formula relating equivalent nominal rates and true rates. Thereafter he requested that the demonstrated formulas assume that the payment and interest periods did indeed "jive." And, he asked the Pack to present the formulas referencing general payment periods without using specific time periods such as years or months. Finally, he asked them to provide examples using the formulas.

Finally, Herkimer asked the Pack to reflect on the fact that the formulas are wonderful for working with nice financial models such as loan payment schedules or bond pricing. However, not all financial scenarios conform to nice models reflecting a series of uniform payments. There are simply times when one has to grind out computations as illustrated in Session #11.

As always, the Pack accepted the challenge of preparing a listing of financial formulas that would be useful in common financial models. They produced the following six sheets and made copies for future reference.

IF i AND $i^{(N)}$ ARE EQUIVALENT RATES, THEN

$$1 + i = \left(1 + \frac{i^{(N)}}{N}\right)^N.$$

SOLVING FOR i : $\quad i = \left(1 + \frac{i^{(N)}}{N}\right)^N - 1.$

EXAMPLE: IF $i^{(12)} = 8\%$, THEN

$$i = \left(1 + \frac{.08}{12}\right)^{12} - 1 = .0829995$$
$$\approx 8.3\%.$$

SOLVING FOR $i^{(N)}$: $\quad 1 + \frac{i^{(N)}}{N} = \left(1 + i\right)^{1/N}$

$$\Rightarrow \quad \frac{i^{(N)}}{N} = \left(1 + i\right)^{1/N} - 1$$

$$\Rightarrow \quad i^{(N)} = N\left[\left(1 + i\right)^{1/N} - 1\right].$$

EXAMPLE: IF $i = 7\%$, THEN

$$i^{(365)} = 365\left[\left(1.07\right)^{1/365} - 1\right]$$
$$= 0.676649$$
$$\approx 6.77\%.$$

REFERENCE: SESSIONS 18-20

234

ACCUMULATION:

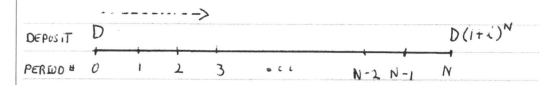

DEPOSIT D $\qquad$ $D(1+i)^N$

PERIOD # 0 1 2 3 ... N-2 N-1 N

EXAMPLE: A DEPOSIT OF $1,000 EARNING AN ANNUAL RATE OF 6.5% WILL ACCUMULATE TO $1000(1.065)^{20} = \$3,523.65$ IN 20 YEARS.

DISCOUNT:

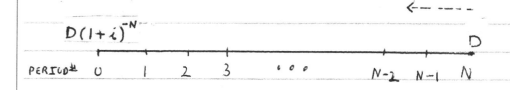

$D(1+i)^{-N}$ $\qquad$ D

PERIOD # 0 1 2 3 ... N-2 N-1 N

EXAMPLE: IF A DEPOSIT EARNS 7.2% AND YOU WISH TO HAVE $50,000 IN 20 YEARS YOU WOULD NEED A SINGLE DEPOSIT OF $50,000(1.072)^{-20}$

$$= \frac{\$50,000}{(1.072)^{20}} = \$12,447.28$$

AT THE PRESENT TIME.

REFERENCE: SESSIONS 3-6

SHEET #3
ACCUMULATION OF MULTIPLE DEPOSITS (i = TRUE RATE FOR PERIOD)

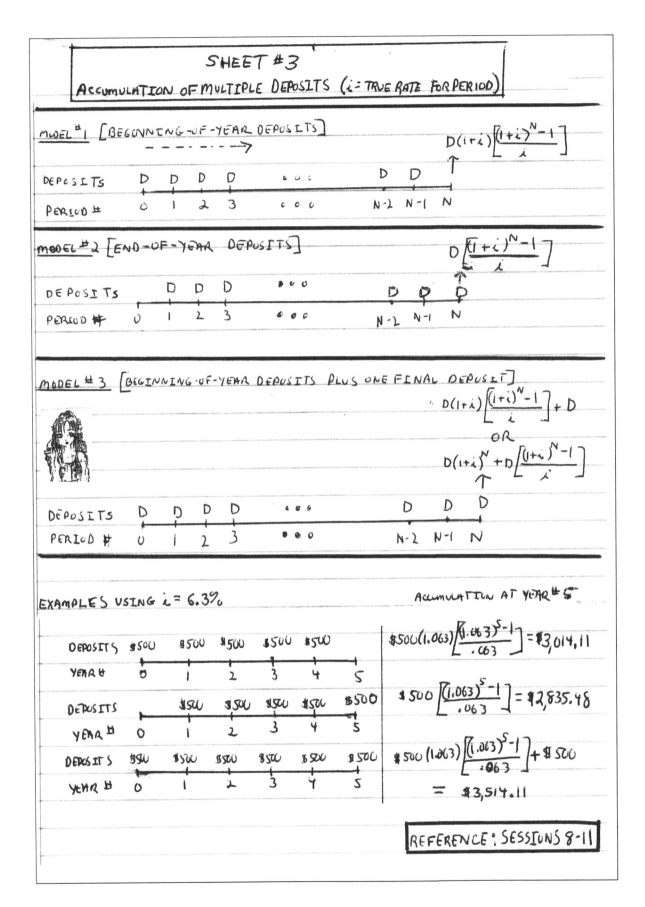

MODEL #1 [BEGINNING-OF-YEAR DEPOSITS]

$$D(1+i)\left[\frac{(1+i)^N-1}{i}\right]$$

DEPOSITS	D	D	D	D	$\circ \circ \circ$		D	D	
PERIOD #	0	1	2	3	$\circ \circ \circ$		N-2	N-1	N

MODEL #2 [END-OF-YEAR DEPOSITS]

$$D\left[\frac{(1+i)^N-1}{i}\right]$$

DEPOSITS		D	D	D	$\bullet \bullet \bullet$		D	D	D
PERIOD #	0	1	2	3	$\bullet \bullet \bullet$		N-2	N-1	N

MODEL #3 [BEGINNING-OF-YEAR DEPOSITS PLUS ONE FINAL DEPOSIT]

$$D(1+i)\left[\frac{(1+i)^N-1}{i}\right] + D$$

OR

$$D(1+i)^N + D\left[\frac{(1+i)^N-1}{i}\right]$$

DEPOSITS	D	D	D	D	$\bullet \bullet \bullet$		D	D	D
PERIOD #	0	1	2	3	$\bullet \bullet \bullet$		N-2	N-1	N

EXAMPLES USING i = 6.3%   ACCUMULATION AT YEAR #5

DEPOSITS	$500	$500	$500	$500	$500	
YEAR #	0	1	2	3	4	5

$$\$500(1.063)\left[\frac{(1.063)^5-1}{.063}\right] = \$3,014.11$$

DEPOSITS		$500	$500	$500	$500	$500
YEAR #	0	1	2	3	4	5

$$\$500\left[\frac{(1.063)^5-1}{.063}\right] = \$2,835.48$$

DEPOSITS	$500	$500	$500	$500	$500	$500
YEAR #	0	1	2	3	4	5

$$\$500(1.063)\left[\frac{(1.063)^5-1}{.063}\right] + \$500$$

$$= \$3,514.11$$

REFERENCE: SESSIONS 8-11

236

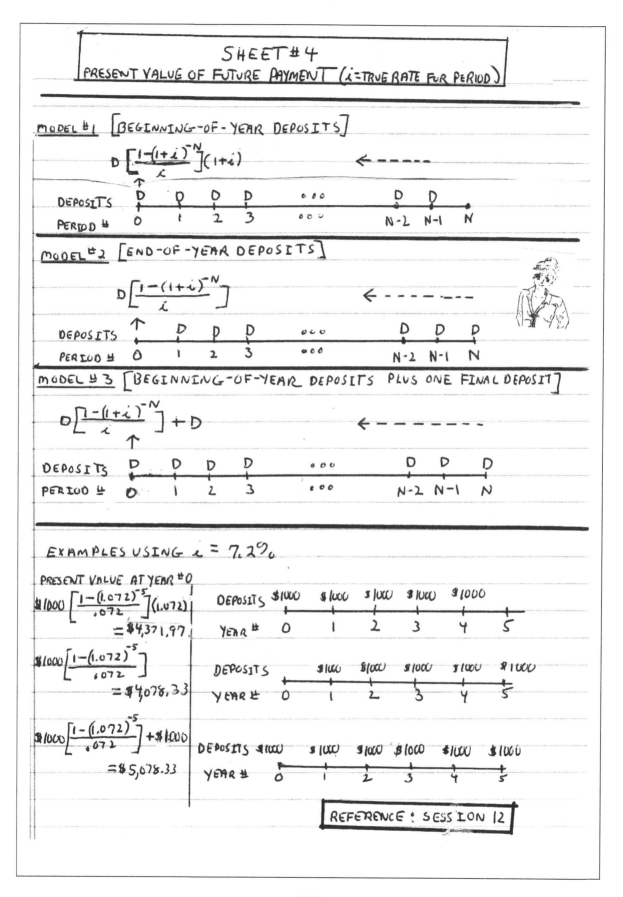

SHEET #4
PRESENT VALUE OF FUTURE PAYMENT (i = TRUE RATE FOR PERIOD)

MODEL #1 [BEGINNING-OF-YEAR DEPOSITS]

$$D\left[\frac{1-(1+i)^{-N}}{i}\right](1+i) \qquad \longleftarrow ----$$

DEPOSITS	D	D	D	D	$\circ\circ\circ$		D	D	
PERIOD #	0	1	2	3	$\circ\circ\circ$		N-2	N-1	N

MODEL #2 [END-OF-YEAR DEPOSITS]

$$D\left[\frac{1-(1+i)^{-N}}{i}\right] \qquad \longleftarrow - - - - - - -$$

DEPOSITS		D	D	D	$\circ\circ\circ$		D	D	D
PERIOD #	0	1	2	3	$\circ\circ\circ$		N-2	N-1	N

MODEL #3 [BEGINNING-OF-YEAR DEPOSITS PLUS ONE FINAL DEPOSIT]

$$D\left[\frac{1-(1+i)^{-N}}{i}\right] + D \qquad \longleftarrow - - - - - -$$

DEPOSITS	D	D	D	D	$\circ\circ\circ$		D	D	D
PERIOD #	0	1	2	3	$\circ\circ\circ$		N-2	N-1	N

EXAMPLES USING $i = 7.2\%$

PRESENT VALUE AT YEAR #0

$$\$1000\left[\frac{1-(1.072)^{-5}}{.072}\right](1.072)$$
$$= \$4,371.97$$

DEPOSITS	$1000	$1000	$1000	$1000	$1000	
YEAR #	0	1	2	3	4	5

$$\$1000\left[\frac{1-(1.072)^{-5}}{.072}\right]$$
$$= \$4,078.33$$

DEPOSITS		$1000	$1000	$1000	$1000	$1000
YEAR #	0	1	2	3	4	5

$$\$1000\left[\frac{1-(1.072)^{-5}}{.072}\right] + \$1000$$
$$= \$5,078.33$$

DEPOSITS	$1000	$1000	$1000	$1000	$1000	$1000
YEAR #	0	1	2	3	4	5

REFERENCE: SESSION 12

237

P = LOAN AMOUNT

N = NUMBER OF PAYMENT PERIODS

i = TRUE

X = PERIODIC PAYMENT

P

THE FORMULA RELATING P, N, i AND X IS

$$P = X\left[\frac{1-(1+i)^{-N}}{i}\right]$$

SOLVING THE EQUATION FOR X YIELDS

$$X = \frac{iP}{1-(1+i)^{-N}}$$

EXAMPLE: IF A 10-YEAR \$50,000 LOAN AT i = 8%
IS REPAID WITH YEAR-END INSTALLMENTS,
THEN EACH OF THE 10 PAYMENTS IS

$$X = \frac{(.08)(\$50,000)}{1-(1.08)^{-10}} = \$7,451.47.$$

EXAMPLE: CONSIDER A 30-YEAR \$400,000 MORTGAGE AT 6%.
IN THIS SITUATION EACH MONTHLY PAYMENT HAS
ASSOCIATED WITH IT A TRUE MONTHLY RATE
OF $\frac{.06}{12}$ = .005, THE MONTHLY INSTALLMENT IS

$$X = \frac{(.005)(\$400,000)}{1-(1.005)^{-360}} = \$2,398.20.$$

REFERENCE: SESSIONS 16, 21

238

M = MATURITY VALUE OF BOND

N = NUMBER OF PERIODS IN TERM OF BOND

i = DESIRED YIELD RATE

y = COUPON RATE

C = yN = VALUE OF PERIODIC COUPON

P = PURCHASE PRICE

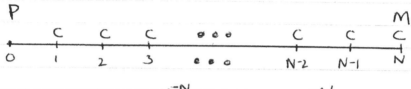

$$P = C\left[\frac{1-(1+i)^{-N}}{i}\right] + M(1+i)^{-N}$$

EXAMPLE: A 20-YEAR \$100,000 BOND WITH
6% ANNUAL COUPONS IS PURCHASED
TO YIELD 7%. THE PURCHASE
PRICE IS

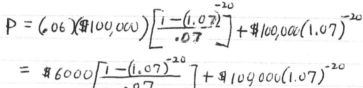

$$P = (.06)(\$100,000)\left[\frac{1-(1.07)^{-20}}{.07}\right] + \$100,000(1.07)^{-20}$$

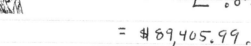

$$= \$6000\left[\frac{1-(1.07)^{-20}}{.07}\right] + \$100000(1.07)^{-20}$$

$$= \$89,405.99.$$

REFERENCE: SESSION 28

239

The Pack felt they had produced sheets that summarized the important formulas they had developed during their many sessions of financial education. Herkimer agreed.

HERKIMER'S FASCINATING FINANCIAL FACTS:

It is no secret that money can be, and has been, a cause of problems for as long as we have used it. This fact even has a Biblical reference.

For the love of money is the root of all evil: which while some coveted after, they have erred from the faith, and pierced themselves through with many sorrows.

First Epistle to Timothy (6:10)

ACTIVITY SET FOR SESSION #43

1. ALCAL activity: Use your calculator to check the computations displayed on the six sheets produced by the Stat Pack during this session.

2. REFLECTION & COMMENT activity:

Many Americans are falling into the trap of buying more than incomes will support because they don't want to wait until they can afford to pay cash. Besides the "I want it now" mentality, people are also finding themselves in debt as the cost of living continues to rise while wages remain the same. When an emergency situation comes up, like a job layoff, an unexpected medical procedure or a natural disaster, the usual outcome is increased credit card debt.

(General note, *Using Credit Cards Wisely*, www.creditrepair.com, 2009)

$$$
$$$

Session #44
HERKIMER AND THE STAT PACK: THE FINAL SESSION

> *Our incomes are like shoes; if too small, they gall and pinch us; but if too large, they cause us to stumble and trip.*
> -John Locke

Pack members knew that Herkimer was with them for a final session on finance. He would soon disappear from their lives as he had previously done after a study of statistics. Perhaps he would return at a later date to work with them on another topic. Then again, perhaps not! The purpose of the last session was not to dwell on sentiment or tearful farewells, but rather to provide a meaningful conclusion to series of sessions on the mathematics of finance.

Herkimer reminded the Pack that they were teenagers and learning about many financial concepts for the first time in their lives. "Someday you will be parents," he said. "Wouldn't it be a good thing if your children learned about money management at a very early age? You certainly wouldn't be able to teach them the mathematics of finance that you have learned in our sessions, but wouldn't it be nice to give them experiences that would allow them to appreciate the advantages of planned savings and how money can actually grow with good investment plans? Let's spend this final session talking about how young children might learn to manage money wisely."

The Pack liked the idea. Herkimer directed the discussion so that it would center around an idea initiated by a Santa Barbara, California woman named Becky Brown who developed strategies involving three piggy banks. One of the pigs was labeled SPENDING, the second was labeled SAVING, and the third was labeled SHARING. "Now Becky presents ideas about how these three pigs can be used to promote money management education for children," he said, "but let's develop our own plan as to how to we might use the pigs with children you may have in the future."

Lots of good thoughts surfaced from this intelligent group of young Pack members. They finally settled on a plan involving the three pigs. They realized there could be many different versions of this plan created by forward-thinking parents, but they wanted one that they could demonstrate with spreadsheet mathematics. Here is the Pack-created plan for a young child they named Herkimer Jr.

(1) *Only quarters would be deposited in the pigs. Herkimer Jr. would be encouraged to save a few quarters each month.*

(2) *The quarters would remain on deposit in the pigs for one year. No withdrawals would be allowed. At the end of one year the money in the SPENDING pig and the SHARING pig could be withdrawn and used for the purposes indicated below. The money in the SAVING pig could not be withdrawn until the end of the second year.*

(3) *SPENDING pig: At the end of each month a penny would be added to the pig for each quarter it contained. At the end of the year Herkimer Jr. could spend the accumulated savings on personal items.*

(4) *SAVING pig: At the end of each month, two cents would be added to the pig for each quarter it contained. As indicated above, this money could not be withdrawn until the end of the second year. (The SAVING pig earns more interest than the SPENDING pig, but money must remain on deposit for a longer period of time before it can be spent.)*

(5) *SHARING pig: At the end of each month, 3 cents would be added to the pig for each quarter it contained. At the end of the year the money would be donated to charity or used to support a worthy humanitarian cause.*

(6) *During the year a set of 25 pennies would NOT be converted to a quarter. Hence at the end of the year the total money represented by the quarters would be the "deposits" made by Herkimer Jr. and the total of the pennies would represent interest earned by the deposits.*

(7) *Early deposits would earn more interest than later deposits. For instance, a quarter invested during the first month of the year would earn 12 cents of interest during the year, whereas one invested in the 12th month would earn just once cent of interest.*

Pack members felt that this plan was certainly one that they could use for children they might have in the future. It does require financial decision making on the part of the investor of money in the pigs. And, it promotes the benefits of planned savings.

Pack teams worked on an interactive spreadsheet program to illustrate a hypothetical year of savings using the three pigs. Their final spreadsheet project follows:

	A	B	C	D	E	F	G	H	I	J
1										
2	SPENDING PIG (Interest = one cent for every quarter in pig at end of month)									
3		Number of	Number of		Total number	Monthly				
4	End of	quarters saved	dollars saved		of quarters in	interest				
5	Month…	during month	during month		SPENDING PIG	payment ($)				
6	Jan	6	$1.50		6	$0.06				
7	Feb	8	$2.00		14	$0.14				
8	Mar	7	$1.75		21	$0.21				
9	Apr	10	$2.50		31	$0.31				
10	May	7	$1.75		38	$0.38				
11	Jun	6	$1.50		44	$0.44				
12	Jul	11	$2.75		55	$0.55				
13	Aug	7	$1.75		62	$0.62				
14	Sep	9	$2.25		71	$0.71				
15	Oct	8	$2.00		79	$0.79				
16	Nov	7	$1.75		86	$0.86				
17	Dec	8	$2.00		94	$0.94				
18										
19			$23.50			$6.01		Total in SPENDING PIG=		$29.51
20										
22	SAVING PIG (Interest = two cents for every quarter in pig at end of month)									
23		Number of	Number of		Total number	Monthly				
24	End of	quarters saved	dollars saved		of quarters in	interest				
25	Month…	during month	during month		SPENDING PIG	payment ($)				
26	Jan	2	$0.50		2	$0.04				
27	Feb	5	$1.25		7	$0.14				
28	Mar	4	$1.00		11	$0.22				
29	Apr	3	$0.75		14	$0.28				
30	May	1	$0.25		15	$0.30				
31	Jun	5	$1.25		20	$0.40				
32	Jul	4	$1.00		24	$0.48				
33	Aug	3	$0.75		27	$0.54				
34	Sep	6	$1.50		33	$0.66				
35	Oct	1	$0.25		34	$0.68				
36	Nov	6	$1.50		40	$0.80				
37	Dec	3	$0.75		43	$0.86				
38										
39			$10.75			$5.40		Total in SAVING PIG=		$16.15
40										
42	SHARING PIG (Interest = three cents for every quarter in pig at end of month)									
43		Number of	Number of		Total number	Monthly				
44	End of	quarters saved	dollars saved		of quarters in	interest				
45	Month…	during month	during month		SPENDING PIG	payment ($)				
46	Jan	3	$0.75		3	$0.09				
47	Feb	2	$0.50		5	$0.15				
48	Mar	3	$0.75		8	$0.24				
49	Apr	3	$0.75		11	$0.33				
50	May	2	$0.50		13	$0.39				
51	Jun	5	$1.25		18	$0.54				
52	Jul	3	$0.75		21	$0.63				
53	Aug	6	$1.50		27	$0.81				
54	Sep	2	$0.50		29	$0.87				
55	Oct	2	$0.50		31	$0.93				
56	Nov	3	$0.75		34	$1.02				
57	Dec	5	$1.25		39	$1.17				
58										
59			$9.75			$7.17		Total in SHARING PIG=		$16.92

Herkimer praised the Pack for a very imaginative and insightful use of Becky Brown's three-piggy bank concept.

Herkimer was about to disappear. He would move on to seek out a group of students who expressed a genuine interest in learning about the power of mathematics. He would present himself as a guide if such a group desired his assistance. He would be visible to those who possessed a genuine desire to learn.

For all others, he would simply be a man who wasn't there.

ACTIVITY SET FOR SESSION #44

1. ALCAL activity: Use your calculator to check the totals displayed on the spreadsheet for each of the three pigs (SPENDING, SHARING, SAVING).

2. REFLECTION & COMMENT activity:

   If you do opt to allow your children to borrow money, you must make it a learning experience. For the very young, introduce the "visibile IOU" --- a glass jar in plain sight, marked IOU. He or she pays back the advance to the jar. Put the IOU jar besides a savings jar, so each time your child makes a payment, he or she is aware that savings won't increase until the borrowed money is returned. And, charge interest. Maybe only pennies, but the borrower will get the idea --- one pays a fee for borrowing money.

   (Elizabeth Lewin and Bernard Ryan, *How to Raise Money-Smart Kids*, www.kidsmoney.org)

3. Suggest other ways one could use Becky Brown's three piggy banks to teach young children concepts of money management. [For more information relating to Becky's piggy banks, visit the internet site http://www.pickpigs.com.]

4. Research or devise other methods for teaching young children about the advantages of saving money.

$$\$$$
$$\$$$

1. A 25-year loan of $200,000 with a true annual interest rate of 7.2% will be repaid with 25 equal end-of-year payments? What is the amount of each payment and the interest portion of the first payment? (SESSION #17)

2. A 30-year $340,000 mortgage has a 6.75% interest rate.
   (a) What is the monthly payment on this mortgage?
   (b) At the end of three years, the outstanding principal on this loan is $328,355. If this is refinanced into a 30-year mortgage at 5.25%, what is the new monthly payment? (SESSION # 38)

3. If the true annual interest rate is 5.9%, what is the accumulation of 20 end-of-year deposits of $15,000 at the end of the 20th year? (SESSION #10)

4. It the true annual interest rate is I, write down the algebraic expression representing the present value of end-of-year payments of K dollars for T years. (SESSION #15)

5. What date appears in Roman numerals at the base of the unfinished pyramid pictured on the back side of a U.S. $1 bill? (SESSION #13)

6. What is the accumulation of a single deposit of $10,000 after 20 years if the annual rate is 6.75% compounded daily? [Recall that this rate can be written $i^{(365)} = 6.75\%$.] (SESSION #18)

7. In a 31-day month, if you saved a quarter the first day, two quarters the second day, four quarters the third day, eight quarters the fourth day, etc. what amount would you be saving on the 31st day? (SESSION #32)

8. A loan of $50,000 is to be repaid with equal year-end installments at the end of each year for the next 10 years. If the annual loan rate is 8.5%, what is the amount of each payment? (SESSION # 16)

9. When did the U.S. government first issue savings bonds and why were they issued? (SESSION #29)

10. If an investment pays an annual rate of 7.86% compounded continuously, what is the annual yield? (SESSION #36)

11. A retirement fund contains $600,000. Yearly amounts of $40,000 will be withdrawn from the fund which is earning a true annual rate of 5.8%. After the first withdrawal, what will be the amount in the fund just prior to the second withdrawal? (SESSION #26)

12. A single deposit of $8,000 earning a true annual rate of 6.5% would accumulate to what amount in 20 years? (SESSION #3)

13. A $40,000 ten-year bond with 6% annual coupons is purchased to yield 7%. What is the purchase price? (SESSION #28)

14. The first credit card that allowed cardholders to create charges at multiple stores was issued in what year? (SESSION #27)

15. How many characters make up the serial number on a U.S. one-dollar bill? How many are numbers and how many are alphabetical letters? (SESSION #1)

16. A single deposit of $100,000 earns interest for 15 years. Find the accumulation at the end of the 15th year if the annual interest rate is (a) a true annual rate of 6%; (b) an annual rate of 6% compounded monthly; (c) an annual rate of 6% compounded continuously. (SESSION #16)

17. How long will it take a deposit to double if the true annual interest rate is 7%? (SESSION #6)

18. A ten-year $100,000 bond with 5% annual coupons is purchased for $94,000. Write the equation that would have to be solved to find the annual yield i. (SESSION #30)

19. What is the displayed time on the Independence Hall clock that appears on the back side of a U.S. one-hundred dollar bill? (SESSION #2)

20. What single deposit earning a true annual rate of 6.8% would be required to accumulate to $100,000 in 20 years? (SESSION #4)

21. A bank advertises that it will pay an annual rate of 5.95% compounded daily on a certificate of deposit. What is the annual yield on this investment? (SESSION #19)

22. What is a *Ponzi scheme*? (SESSION #31)

23. Use the *Rule of 72* to estimate how long it would take a single deposit to double at a true annual rate of 6.83%. Then calculate the actual time required. (SESSION #33)

24. It the true annual interest rate is 6.4%, what single deposit would one have to make now in order to be able to withdraw $20,000 at the end of each year for the next 10 years? (SESSION #12)

25. A 20-year $100,000 bond with 6% semiannual coupons is purchased to yield an annual rate of 8% compounded semiannually. What is (a) the purchase price? (b) the book value after the first coupon has been paid? (SESSION #39)

$$\$$$

26. The interest rate on a credit card is 26% compounded monthly. If the outstanding balance is $2,940 what is the interest portion of the next due payment? (SESSION #24)

27. I deposit $10,000 now, $20,000 two years from now, and $40,000 five years from now. If these deposits earn a true annual rate of 5.98%, what is the accumulation of these deposits ten years from now? (SESSION #27)

28. Upon retiring a retiree wants to start withdrawing annual amounts from a retirement fund. If the first annual withdrawal is $40,000 and following withdrawals increase to compensate for a 4% inflation rate, what are the first four annual amounts withdrawn from the fund? (SESSION #40)

29. What nominal annual rate $i^{(12)}$ is *equivalent* to an annual yield i = 13%? (SESSION #20)

30. Suppose you have an outstanding amount of $623 on a credit card with an annual rate of 22% compounded monthly. If you make a minimum payment of $15 at the next bill, how much of this payment is interest? (SESSION #25)

$$\$$$

31. A loan of $60,000 will be repaid with monthly payments for 10 years. If the loan interest rate is $i^{(12)} = 14\%$, what is the amount of the monthly payment? (SESSION #21)

32. What deposit earning a true annual rate of 7% would be required to withdraw payments of $1,000 at the end of each year for 12 years? (SESSION #14)

33. A *payday loan* agreement allows one to borrow $500 now in exchange for repaying the loan with an amount of $580 in two weeks. What is the true annual rate on this loan? (SESSION #34)

34. A credit card balance of $6,500 is transferred to another credit card offering a teaser rate of 0% for the first three months. Thereafter the rate becomes $i^{(12)} = 22\%$. If you make a monthly payment of $125, what portion of the 4th payment is interest? (SESSION #41)

35. A $480,000 thirty-year mortgage has a rate of 6.75%. What is the amount of each monthly payment? (SESSION #22)

$$\$$$

36. What equal beginning-of-year deposits would be required to accumulate to $25,000 in 15 years if the true annual interest rate is 7%? (SESSION #9)

37. A 20-year-old woman invests $5,000 now and plans to let it earn interest until she reaches age 65. A 40-year-old woman invests $15,000 now and plans to let it earn interest until she reaches age 65. It the money earns a true annual rate of 7%, what is the accumulated value of each when the individuals reach age 65? (SESSION #35)

38. A thirty-year $320,000 adjustable rate mortgage (ARM) has an initial rate of 5.4%. After 1 year, the loan rate changes to 6.4%.
    (a) What is the monthly payment for the first year?
    (b) What is the principal outstanding after the first year?
    (c) What is the revised monthly payment after the first year?
    (SESSION #42)

39. If $5,000 accumulates to $12,000 in 7 years, what is the annual interest rate? (SESSION #5)

40. If the true annual interest rate is 6%, what is the accumulation of 20 beginning-of-year deposits of $5,000 at the end of the twentieth year? (SESSION #8)

$$\$$$

41. What is a primary advantage of making extra non-required payments on a mortgage? (SESSION #37)

42. At a 3.6% inflation rate, an item costing $100 now will cost what amount in 10 years. (SESSION #7)

43. A 30-year $300,000 mortgage is desired. What are the monthly payments if the mortgage interest rate is (a) 5.7%? (b) 6.7%? (SESSION #23)

44. How much interest would be earned by a single deposit of $10,000 earning a true annual rate of 10% for 15 years? (SESSION #3)

45. A twenty-year $290,000 adjustable rate mortgage (ARM) has an initial rate of 4.9%. After 5 years, the loan rate changes to 6.2%.
   (a) What is the monthly payment for the first 5 years?
   (b) What is the principal outstanding after the first 5 years?
   (c) What is the revised monthly payment after the first 5 years?
   (SESSION #42)

$$\$$$

46. A single deposit of $50,000 earns a true annual rate of 6% for five years. The rate is then increased to 7% for the next five years, and then to 8% for the final five years. What is the accumulation of this investment at the end of the 15th year? (SESSION #11)

47. If the face value of a Series E bond is $500, then the purchase price is $375. If this bond could be redeemed in 6 years and 3 months, what is the annual yield for the purchaser? (SESSION #29)

48. At a true annual interest rate of 12.6%, how long would it take a deposit of $1,000 to accumulate to $8,000? (SESSION #4)

49. It the true annual interest rate is I, write down the algebraic expression representing the accumulated value of end-of-year payments of K dollars for T years at the end of the year #T. (SESSION #15)

50. If one deposits $50,000 now that earns a true annual rate of 6%, what equal withdrawals could be made at the end of each year for the next 20 years? (SESSION #12)

$$\$$$

51. A $400,000 loan is repaid with monthly payments over a period of 15 years. It the loan rate is i(12) = 9%, what is the total amount of interest paid over the 15-year period. (SESSION #21)

52. A credit card has an annual rate of 25% compounded monthly. If your outstanding balance is $586 and you make a payment of $15 at the next billing, what is the outstanding balance after this payment? (SESSION #25)

53. If the true annual interest rate is 8%, what deposit will accumulate to $100,000 in 20 years? (SESSION #6)

54. A retirement fund contains $800,000 and earns a true annual interest rate of 6%. If annual withdrawals of $40,000 will be made, why will this fund never run out of money? (SESSION #26)

250

55. Forty beginning-of-year deposits of $1,000 earning a true annual rate of 7% will accumulate to what amount at the end of the 40th year? (SESSION #8)

$$$$$$$$$$$$$$$$$$$$$$$$$$$$$$$$

56. If a $1 bill is torn into 2 pieces and one of the pieces is lost, is the other piece spendable? (SESSION #40)

57. A 20-year $1,000,000 bond with 8% annual coupons is purchased to yield 6%. What is the purchase price? (SESSION #28)

58. What is the total of all payments on a twenty-year 6.9% mortgage loan of $500,000? (SESSION #22)

59. A 15-year $100,000 bond with 7% semiannual coupons is purchased to yield an annual rate of 5% compounded semiannually. What is (a) the purchase price? (b) the book value after the first coupon has been paid? (SESSION #39)

60. A $300,000 mortgage is desired. The loan rate is 6%. What are the monthly payments if the term of the mortgage is (a) 20 years? (b) 30 years? (SESSION #23)

$$$$$$$$$$$$$$$$$$$$$$$$$$$$$$

61. Consider a 30-year $420,000 mortgage. What will be the total of all monthly payment over the term of the loan if the interest rate is (a) 7.2%? (b) 5.2%? (SESSION #38)

62. A payday loan of $100 requires a repayment of $125 in 2 weeks. What is the true annual interest rate on this loan? (SESSION #34)

63. A single deposit of $100,000 earns a true annual rate of 10% for 10 years, an annual rate $i^{(12)} = 10\%$ for the next 5 years, and an annual rate $i^{(365)} = 10\%$ for the next 10 years. What is the accumulation of this investment at the end of the 25th year? (SESSION #19)

64. If an investment increases 4-fold in 17 years, what is the annual interest rate earned? (SESSION #5)

65. A ten-year $25,000 bond with 6% annual coupons is purchased for $24,000. If i represents the annual yield for the purchaser, write the equation that would need to be solved to find the yield? (SESSION #30)

$$$$$$$$$$$$$$$$$$$$$$$$$$$$$$$$

66. A $700,000 retirement fund earns an annual rate $i^{(12)} = 6\%$. A retiree wants to withdraw monthly payments of $3,500. What amount will be in the fund just before the second withdrawal? (SESSION #27)

67. If one deposits $1,000 at the end of each year for 20 years at a true rate of 7%, what would be the accumulated value of this investment at the end of the 20th year? (SESSION #14)

68. A loan of $20,000 is being repaid with equal year-end payments for 5 years. If the true annual interest rate on this loan is 15%, what is the total amount of interest paid over the 5-year period? (SESSION #16)

69. A credit card balance of $8,100 is transferred to another credit card offering a teaser rate of 0% for the first six months. Thereafter the rate becomes $i^{(12)} = 26\%$. If you make a monthly payment of $200, what are the interest and principal portions of the 7th payment? (SESSION #41)

70. Consider the five deposits displayed below and assume they earn interest at a true annual rate of 6%.

Deposit			$5,000	$5,000	$5,000	$5000	$5,000		
Year	0	1	2	3	4	5	6	7	8

    (a) What is the accumulated value of the deposits at year #8?
    (b) What is the present value of the deposits at year #0?
    (c) How are the answers in (a) and (b) related? If you knew answer (a), how could you easily obtain answer (b)?
(SESSION #15)

$$\$$$

71. A 25-year loan of $200,000 with a true annual interest rate of 7.2% will be repaid with 25 equal end-of-year payments? What is the amount of the outstanding principal after the first payment? (SESSION #17)

72. Who is the only female to have her portrait on U.S. paper money? (SESSION #5)

73. The outstanding balance on a credit card is $1,960. If the credit card rate is 23% and a payment of $40 is made at the next due date, what portion of this payment goes to reducing the amount owed? (SESSION #24)

74. A deposit of $20,000 now will allow you to withdraw $30,000 in 6 years and $50,000 in 10 years. If x is the true annual interest rate earned on this investment, what equation must be solved to find the value of x? (SESSION #13)

75. Assume that a $280,000 mortgage requires a monthly payment of $1,450. For the 17th payment, a check for $1,750 is submitted. What effect will this have on the outstanding balance after the payment is recorded and applied to the loan schedule? (SESSION #37)

76. If an investment ad states that invested money earns an annual rate of 12% compounded monthly, what is the annual yield on such an investment? (SESSION #19)

77. Consider five beginning-of-year deposits of $50,000.

	$50,000	$50,000	$50,000	$50,000	$50,000	.
Year	0	1	2	3	4	5

What is the accumulation of these deposits at the end of the fifth year if the interest rate is (a) a true annual rate of 6.5%? (b) an annual rate of 6.5% compounded continuously? (SESSION #36)

78. Write the equation that would have to be solved to find the annual yield i on a 25-year $1,000,000 bond with 6% annual coupons purchased for $1,150,000. (SESSION #30)

79. If the annual interest rate is 8.8% compounded monthly, what single deposit must be made now to accumulate to $30,000 in 20 years? (SESSION #18)

80. If the inflation rate is 4.2%, an item worth $100 now will have what value in 12 years. (SESSION #7)

81. If 10 beginning-of-year deposits of $1,000 accumulate to $21,000 at the end of the 10th year, write the equation that would have to be solved to find the true annual rate x. (SESSION #9)

82. A credit card has a balance of $8,600. What amount of interest would be required in the next payment if the interest rate is (a) $i^{(12)} = 8\%$? (b) $i^{(12)} = 18\%$? (c) $i^{(12)} = 28\%$? (SESSION #24)

83. What nominal annual rate $i^{(365)}$ is equivalent to an annual yield $i = 20\%$? (SESSION #20)

84. What is the accumulation of the displayed deposits at the end of the 5th year if the true annual interest rate is 11%? (SESSION #10)

	$5000	$5000	$5000	$5000	$5000	$5000
Year	0	1	2	3	4	5

85. If you want to have $100,000 available in 20 years, what single deposit must be made now to accumulate to that amount at (a) a true annual rate of 9%? (b) an annual rate $i^{(12)} = 9\%$? (c) an annual rate $i^{(365)} = 9\%$? (SESSION #19)

$$\$$$

86. Assuming a true annual interest rate of 7%, which investment will be worth more at age 65? For each investment, calculate the total amount of money invested and the accumulated value of the investment at age 65.

   Investment #1: Starting at age 20 an individual makes five beginning-of-year deposits of $5,000 and then simply lets the money accumulate without making any more deposits.

   Investment #2: Starting at age 50 an individual makes fifteen beginning-of-year deposits of $10,000.

   (SESSION #35)

87. Name the two men featured on U.S. currency who were never Presidents of the United States? (SESSION #17)

88. An individual borrows $10,000 and has the option of repaying the loan with
   (A) a single payment of $20,000 in 5 years, or
   (B) a single payment of $45,000 in 10 years.
   In terms of annual yield, which choice is more favorable to the borrower?
   (SESSION #5)

89. A twenty-year $100,000 bond with 6% annual coupons is sold just after the owner receives the 15th coupon. If it is purchased to yield 10% annually, what is the purchase price? (SESSION #28)

90. Consider a thirty-year $250,000 mortgage with a 6.25% interest rate.
   (a) What is the monthly payment?
   (b) What is the principal outstanding after the first payment?
   (c) What is the interest portion of the second payment?
   (d) If the borrower pays an extra $5,000 in the first payment, what is the interest portion of the second payment?
   (SESSION #37)

$$\$$$

91. If deposits earn an annual rate of 6.2% compounded daily, what single deposit made now will allow for a withdrawal of $10,000 in five years and a second withdrawal of $10,000 in 10 years? (SESSION #19)

92. A payday-style loan requires that for each $100 borrowed, the borrower must repay $150 in one month. What is the annual yield on this loan? (SESSION # 34)

93. If an investment of $1,000 accumulates to $6,000 in 19 years, what is the annual yield? (SESSION #5)

94. A $470,000 mortgage has a 5.75% interest rate. Calculate the total of all payments on this mortgage if the term is (a) 30 years; (b) 15 years. (SESSION #22)

95. To repay a loan of $8,000 the borrower agrees to pay the lender $4,000 in three years and another payment of $7,000 in five years. If x is the annual yield on this loan, write the equation that would have to be solved to find x. (SESSION #13)

$$\$$$

96. Which rate produces the greatest annual yield, $i^{(2)} = 8.12\%$ or $i^{(365)} = 7.98\%$? (SESSION # 18)

97. What U.S. President is featured on the not-often-seen $2 bill? (SESSION #36)

98. Starting now at age 65 a retiree withdraws annual payments of $50,000 from a retirement fund. Assuming a 2.5% inflation rate, what will the withdrawal at age 80 be worth in terms of presently-priced items? (SESSION #40)

99. A Series EE savings bond could be purchased for 50% of its face value. If a $500 Series EE bond can be redeemed in 15 years and 9 months, what is the annual yield on this investment? (SESSION #29)

100. If the true annual interest rate is 6.2%, what is the accumulation of the five displayed deposits at year #10? (SESSION #11)

Deposit	$1,000	$2,000	$3,000	$4,000	$5,000						
Year #	0	1	2	3	4	5	6	7	8	9	10

$$\$$$

APPENDIXES

APPENDIX A
Algebraic Derivation of Formulas

APPENDIX B

First table:
A complete 30-year mortgage payment schedule

Second table:
A complete credit card payment schedule with minimum monthly payments.

APPENDIX C
Solutions and Comments for Selected Activities

APPENDIX A
Algebraic Derivation of Formulas

Session #3 Formulas

Let A_n = accumulated value of a single deposit of P dollars at the end of the n-th year at rate i per year. Then

$A_0 = P$

$A_1 = P + iP = P(1 + i)$

$A_2 = P(1 + i) + iP(1 + i) = P(1 + i)(1 + i) = P(1 + i)^2$

$A_3 = P(1 + i)^2 + iP(1 + i)^2 = P(1 + i)^2(1 + i) = P(1 + i)^3$

....

$A_n = P(1 + i)^{n-1} + iP(1 + i)^{n-1} = P(1 + i)^{n-1}(1 + i) = P(1 + i)^n$

The interest earned after n years is

$P(1 + i)^n - P = P[(1 + i)^n - 1]$

Session #4 Formulas

Let D = the single deposit required to accumulate to A dollars at the end of the n-th year at rate i per year. Then

$$D(1 + i)^n = A ==> D = A/(1+i)^n = A(1 + i)^{-n}$$

The interest earned is

$A - A(1 + i)^{-n}$

Session #8 Formulas

A series of the form $a + ar + ar^2 + ar^3 \ldots + ar^{n-1}$ is a **geometric series** with n terms. If r is not 1, then there is a convenient formula for the sum of the series which will be derived here:

Let $\quad S_n = a + ar + ar^2 + ar^3 \ldots + ar^{n-2} + ar^{n-1}$

Then $\quad rS_n = \quad ar + ar^2 + ar^3 \ldots + ar^{n-2} + ar^{n-1} + ar^n$

Subtracting the two members of the first equation from those in the second yields

$$rS_n - S_n = S_n(r - 1) = ar^n - a = a(r^n - 1) ==> S_n = a(r^n - 1)/(r - 1)$$

In Session #8, this diagram appears.

							A
Deposit	D	D	D	------	D	D	.
Y = End of Year	0	1	2	------	n-2	n-1	n

The accumulation of the displayed deposits at annual rate i is

$$\mathbf{A} = D(1+i) + D(1+i)^2 + \ldots + D(1+i)^{n-1} + D(a+i)^n$$
$$= D(1+i)[1 + (1+i) + (1+i)^2 + \ldots + (1+i)^{n-1}]$$

Within the brackets is a geometric series containing n terms with $a = 1$ and $r = (1+i)$. Using the geometric sum formula developed above, the sum of the bracketed series is $1[(1+i)^n - 1]/[(1+i) - 1] = [(1+i)^n - 1]/i$.

Hence $\mathbf{A} = D(1+i)\ [(1+i)^n - 1]/i$.

Session #10 Formulas

In Session #10, this diagram appears.

							A
Deposit		D	D	------	D	D	D.
Y = End of Year	0	1	2	------	n-2	n-1	n

The accumulation of the displayed deposits at annual rate i is

$$\mathbf{A} = D + D(1+i) + D(1+i)^2 + \ldots + D(1+i)^{n-2} + D(a+i)^{n-1}$$
$$= D[1 + (1+i) + (1+i)^2 + \ldots + (1+i)^{n-1}]$$

Within the brackets is a geometric series containing n terms with $a = 1$ and $r = (1+i)$. Using the geometric sum formula developed above, the sum of the bracketed series is $1[(1+i)^n - 1]/[(1+i) - 1] = [(1+i)^n - 1]/i$.

Hence $\mathbf{A} = D[(1+i)^n - 1]/i$.

Sesson #12 Formulas

In Session #12, this diagram appears.

	P						
Deposit		D	D	------	D	D	D.
Y = End of Year	0	1	2	------	n-2	n-1	n

If the annual interest rate is i the single deposit P (present value) required to withdraw the displayed year-end payments D is

$$\mathbf{P} = D/(1+i) + D/(1+i)^2 + \ldots + D/(1+i)^{n-2} + D/(1+i)^{n-1} + D/(1+i)^n$$

Letting $v = 1/(1+i) = (1+i)^{-1}$, we have

$$\mathbf{P} = Dv + Dv^2 + \ldots + Dv^{n-2} + Dv^{n-1} + Dv^n = Dv(1 + v + v^2 + \ldots + v^{n-2} + v^{n-1})$$

Using the formula for the sum of a geometric series, we have

$$P = Dv(1-v^n)/(1-v) = D(1/(1+i))(1-v^n)/(1-v) = D(1-v^n)/[(1+i)-1]$$
$$= D(1-v^n)/i = D[1-(1+i)^{-n}]/i.$$

Session #17 Formulas

After 1 year, payment = P, interest required = iA.

Principal in first payment = $P - iA = P(1 - iA/P) = P[1 - iA/(iA/(1 - (1+i)^{-n}))]$
$= P[1 - 1 + (1+i)^{-n}] = P(1+i)^{-n} = P/(1+i)^n.$

The principal outstanding after the first payment is the present value of the remaining n-1 payments. This value is $P[1-(1+i)^{-(n-1)}]/i$.

The principal in the second payment is $P - iP[1-(1+i)^{-(n-1)}]/i = P - P[1-(1+i)^{-(n-1)}]$
$= P(1+i)^{-(n-1)} = P/(1+i)^{n-1}.$

Continuing this process, one can determine that the principal in payment #k is $P/(1+i)^{n+1-k}$.

Appendix B

First table: A complete 30-year mortgage payment schedule

Second table: A complete credit card payment schedule with minimum monthly payments.

COMPLETE 30-YEAR MORTGAGE PAYMENT SCHEDULE (Session #23)

MORTGAGE SPREADSHEET...constructed by Herkimer's Stat Pack
User inputs LOAN AMOUNT, INTEREST RATE, and NUMBER OF YEARS (TERM OF MORTGAGE).

$300,000	**<---INPUT LOAN AMOUNT**
	<---INPUT INTEREST RATE (Annual Rate Compounded
6.00%	**Monthly)**
30	**<---INPUT NUMBER OF YEARS (TERM OF MORTGAGE)**

$1,798.65	<----MONTHLY PAYMENT (Calculated)

Month #	Payment	Interest	Principal	Princ. Outstand.	% Interest in payment	
0				$300,000.00		
1	$1,798.65	$1,500.00	$298.65	$299,701.35	83.40%	
2	$1,798.65	$1,498.51	$300.14	$299,401.20	83.31%	
3	$1,798.65	$1,497.01	$301.65	$299,099.56	83.23%	
4	$1,798.65	$1,495.50	$303.15	$298,796.40	83.15%	
5	$1,798.65	$1,493.98	$304.67	$298,491.73	83.06%	
6	$1,798.65	$1,492.46	$306.19	$298,185.54	82.98%	
7	$1,798.65	$1,490.93	$307.72	$297,877.82	82.89%	
8	$1,798.65	$1,489.39	$309.26	$297,568.56	82.81%	
9	$1,798.65	$1,487.84	$310.81	$297,257.75	82.72%	
10	$1,798.65	$1,486.29	$312.36	$296,945.38	82.63%	
11	$1,798.65	$1,484.73	$313.92	$296,631.46	82.55%	
12	$1,798.65	$1,483.16	$315.49	$296,315.96	82.46%	1 year
13	$1,798.65	$1,481.58	$317.07	$295,998.89	82.37%	
14	$1,798.65	$1,479.99	$318.66	$295,680.24	82.28%	
15	$1,798.65	$1,478.40	$320.25	$295,359.99	82.19%	
16	$1,798.65	$1,476.80	$321.85	$295,038.13	82.11%	
17	$1,798.65	$1,475.19	$323.46	$294,714.67	82.02%	
18	$1,798.65	$1,473.57	$325.08	$294,389.59	81.93%	
19	$1,798.65	$1,471.95	$326.70	$294,062.89	81.84%	
20	$1,798.65	$1,470.31	$328.34	$293,734.55	81.75%	
21	$1,798.65	$1,468.67	$329.98	$293,404.58	81.65%	
22	$1,798.65	$1,467.02	$331.63	$293,072.95	81.56%	
23	$1,798.65	$1,465.36	$333.29	$292,739.66	81.47%	
24	$1,798.65	$1,463.70	$334.95	$292,404.71	81.38%	2 years
25	$1,798.65	$1,462.02	$336.63	$292,068.08	81.28%	
26	$1,798.65	$1,460.34	$338.31	$291,729.77	81.19%	

27	$1,798.65	$1,458.65	$340.00	$291,389.76	81.10%
28	$1,798.65	$1,456.95	$341.70	$291,048.06	81.00%
29	$1,798.65	$1,455.24	$343.41	$290,704.65	80.91%
30	$1,798.65	$1,453.52	$345.13	$290,359.52	80.81%
31	$1,798.65	$1,451.80	$346.85	$290,012.67	80.72%
32	$1,798.65	$1,450.06	$348.59	$289,664.08	80.62%
33	$1,798.65	$1,448.32	$350.33	$289,313.75	80.52%
34	$1,798.65	$1,446.57	$352.08	$288,961.67	80.43%
35	$1,798.65	$1,444.81	$353.84	$288,607.82	80.33%
36	$1,798.65	$1,443.04	$355.61	$288,252.21	80.23% 3 years
37	$1,798.65	$1,441.26	$357.39	$287,894.82	80.13%
38	$1,798.65	$1,439.47	$359.18	$287,535.64	80.03%
39	$1,798.65	$1,437.68	$360.97	$287,174.67	79.93%
40	$1,798.65	$1,435.87	$362.78	$286,811.89	79.83%
41	$1,798.65	$1,434.06	$364.59	$286,447.30	79.73%
42	$1,798.65	$1,432.24	$366.42	$286,080.88	79.63%
43	$1,798.65	$1,430.40	$368.25	$285,712.64	79.53%
44	$1,798.65	$1,428.56	$370.09	$285,342.55	79.42%
45	$1,798.65	$1,426.71	$371.94	$284,970.61	79.32%
46	$1,798.65	$1,424.85	$373.80	$284,596.81	79.22%
47	$1,798.65	$1,422.98	$375.67	$284,221.14	79.11%
48	$1,798.65	$1,421.11	$377.55	$283,843.60	79.01% 4 years
49	$1,798.65	$1,419.22	$379.43	$283,464.16	78.90%
50	$1,798.65	$1,417.32	$381.33	$283,082.83	78.80%
51	$1,798.65	$1,415.41	$383.24	$282,699.60	78.69%
52	$1,798.65	$1,413.50	$385.15	$282,314.44	78.59%
53	$1,798.65	$1,411.57	$387.08	$281,927.36	78.48%
54	$1,798.65	$1,409.64	$389.01	$281,538.35	78.37%
55	$1,798.65	$1,407.69	$390.96	$281,147.39	78.26%
56	$1,798.65	$1,405.74	$392.91	$280,754.47	78.16%
57	$1,798.65	$1,403.77	$394.88	$280,359.59	78.05%
58	$1,798.65	$1,401.80	$396.85	$279,962.74	77.94%
59	$1,798.65	$1,399.81	$398.84	$279,563.90	77.83%
60	$1,798.65	$1,397.82	$400.83	$279,163.07	77.71% 5 years
61	$1,798.65	$1,395.82	$402.84	$278,760.23	77.60%
62	$1,798.65	$1,393.80	$404.85	$278,355.38	77.49%
63	$1,798.65	$1,391.78	$406.87	$277,948.51	77.38%
64	$1,798.65	$1,389.74	$408.91	$277,539.60	77.27%
65	$1,798.65	$1,387.70	$410.95	$277,128.65	77.15%
66	$1,798.65	$1,385.64	$413.01	$276,715.64	77.04%
67	$1,798.65	$1,383.58	$415.07	$276,300.56	76.92%
68	$1,798.65	$1,381.50	$417.15	$275,883.42	76.81%
69	$1,798.65	$1,379.42	$419.23	$275,464.18	76.69%
70	$1,798.65	$1,377.32	$421.33	$275,042.85	76.58%
71	$1,798.65	$1,375.21	$423.44	$274,619.41	76.46%
72	$1,798.65	$1,373.10	$425.55	$274,193.86	76.34% 6 years
73	$1,798.65	$1,370.97	$427.68	$273,766.18	76.22%

74	$1,798.65	$1,368.83	$429.82	$273,336.36	76.10%
75	$1,798.65	$1,366.68	$431.97	$272,904.39	75.98%
76	$1,798.65	$1,364.52	$434.13	$272,470.26	75.86%
77	$1,798.65	$1,362.35	$436.30	$272,033.96	75.74%
78	$1,798.65	$1,360.17	$438.48	$271,595.47	75.62%
79	$1,798.65	$1,357.98	$440.67	$271,154.80	75.50%
80	$1,798.65	$1,355.77	$442.88	$270,711.92	75.38%
81	$1,798.65	$1,353.56	$445.09	$270,266.83	75.25%
82	$1,798.65	$1,351.33	$447.32	$269,819.51	75.13%
83	$1,798.65	$1,349.10	$449.55	$269,369.96	75.01%
84	$1,798.65	$1,346.85	$451.80	$268,918.16	74.88% 7 years
85	$1,798.65	$1,344.59	$454.06	$268,464.10	74.76%
86	$1,798.65	$1,342.32	$456.33	$268,007.77	74.63%
87	$1,798.65	$1,340.04	$458.61	$267,549.15	74.50%
88	$1,798.65	$1,337.75	$460.91	$267,088.25	74.37%
89	$1,798.65	$1,335.44	$463.21	$266,625.04	74.25%
90	$1,798.65	$1,333.13	$465.53	$266,159.51	74.12%
91	$1,798.65	$1,330.80	$467.85	$265,691.66	73.99%
92	$1,798.65	$1,328.46	$470.19	$265,221.46	73.86%
93	$1,798.65	$1,326.11	$472.54	$264,748.92	73.73%
94	$1,798.65	$1,323.74	$474.91	$264,274.01	73.60%
95	$1,798.65	$1,321.37	$477.28	$263,796.73	73.46%
96	$1,798.65	$1,318.98	$479.67	$263,317.06	73.33% 8 years
97	$1,798.65	$1,316.59	$482.07	$262,835.00	73.20%
98	$1,798.65	$1,314.17	$484.48	$262,350.52	73.06%
99	$1,798.65	$1,311.75	$486.90	$261,863.62	72.93%
100	$1,798.65	$1,309.32	$489.33	$261,374.29	72.79%
101	$1,798.65	$1,306.87	$491.78	$260,882.51	72.66%
102	$1,798.65	$1,304.41	$494.24	$260,388.27	72.52%
103	$1,798.65	$1,301.94	$496.71	$259,891.56	72.38%
104	$1,798.65	$1,299.46	$499.19	$259,392.36	72.25%
105	$1,798.65	$1,296.96	$501.69	$258,890.67	72.11%
106	$1,798.65	$1,294.45	$504.20	$258,386.48	71.97%
107	$1,798.65	$1,291.93	$506.72	$257,879.76	71.83%
108	$1,798.65	$1,289.40	$509.25	$257,370.50	71.69% 9 years
109	$1,798.65	$1,286.85	$511.80	$256,858.71	71.55%
110	$1,798.65	$1,284.29	$514.36	$256,344.35	71.40%
111	$1,798.65	$1,281.72	$516.93	$255,827.42	71.26%
112	$1,798.65	$1,279.14	$519.51	$255,307.90	71.12%
113	$1,798.65	$1,276.54	$522.11	$254,785.79	70.97%
114	$1,798.65	$1,273.93	$524.72	$254,261.07	70.83%
115	$1,798.65	$1,271.31	$527.35	$253,733.72	70.68%
116	$1,798.65	$1,268.67	$529.98	$253,203.74	70.53%
117	$1,798.65	$1,266.02	$532.63	$252,671.11	70.39%
118	$1,798.65	$1,263.36	$535.30	$252,135.81	70.24%
119	$1,798.65	$1,260.68	$537.97	$251,597.84	70.09%
120	$1,798.65	$1,257.99	$540.66	$251,057.17	69.94% 10 years

121	$1,798.65	$1,255.29	$543.37	$250,513.81	69.79%
122	$1,798.65	$1,252.57	$546.08	$249,967.73	69.64%
123	$1,798.65	$1,249.84	$548.81	$249,418.91	69.49%
124	$1,798.65	$1,247.09	$551.56	$248,867.36	69.33%
125	$1,798.65	$1,244.34	$554.31	$248,313.04	69.18%
126	$1,798.65	$1,241.57	$557.09	$247,755.96	69.03%
127	$1,798.65	$1,238.78	$559.87	$247,196.08	68.87%
128	$1,798.65	$1,235.98	$562.67	$246,633.41	68.72%
129	$1,798.65	$1,233.17	$565.48	$246,067.93	68.56%
130	$1,798.65	$1,230.34	$568.31	$245,499.62	68.40%
131	$1,798.65	$1,227.50	$571.15	$244,928.46	68.25%
132	$1,798.65	$1,224.64	$574.01	$244,354.45	68.09% 11 years
133	$1,798.65	$1,221.77	$576.88	$243,777.57	67.93%
134	$1,798.65	$1,218.89	$579.76	$243,197.81	67.77%
135	$1,798.65	$1,215.99	$582.66	$242,615.15	67.61%
136	$1,798.65	$1,213.08	$585.58	$242,029.57	67.44%
137	$1,798.65	$1,210.15	$588.50	$241,441.07	67.28%
138	$1,798.65	$1,207.21	$591.45	$240,849.62	67.12%
139	$1,798.65	$1,204.25	$594.40	$240,255.22	66.95%
140	$1,798.65	$1,201.28	$597.38	$239,657.84	66.79%
141	$1,798.65	$1,198.29	$600.36	$239,057.48	66.62%
142	$1,798.65	$1,195.29	$603.36	$238,454.12	66.45%
143	$1,798.65	$1,192.27	$606.38	$237,847.74	66.29%
144	$1,798.65	$1,189.24	$609.41	$237,238.32	66.12% 12 years
145	$1,798.65	$1,186.19	$612.46	$236,625.86	65.95%
146	$1,798.65	$1,183.13	$615.52	$236,010.34	65.78%
147	$1,798.65	$1,180.05	$618.60	$235,391.74	65.61%
148	$1,798.65	$1,176.96	$621.69	$234,770.05	65.44%
149	$1,798.65	$1,173.85	$624.80	$234,145.25	65.26%
150	$1,798.65	$1,170.73	$627.93	$233,517.32	65.09%
151	$1,798.65	$1,167.59	$631.06	$232,886.26	64.91%
152	$1,798.65	$1,164.43	$634.22	$232,252.04	64.74%
153	$1,798.65	$1,161.26	$637.39	$231,614.64	64.56%
154	$1,798.65	$1,158.07	$640.58	$230,974.07	64.39%
155	$1,798.65	$1,154.87	$643.78	$230,330.28	64.21%
156	$1,798.65	$1,151.65	$647.00	$229,683.28	64.03% 13 years
157	$1,798.65	$1,148.42	$650.24	$229,033.05	63.85%
158	$1,798.65	$1,145.17	$653.49	$228,379.56	63.67%
159	$1,798.65	$1,141.90	$656.75	$227,722.81	63.49%
160	$1,798.65	$1,138.61	$660.04	$227,062.77	63.30%
161	$1,798.65	$1,135.31	$663.34	$226,399.43	63.12%
162	$1,798.65	$1,132.00	$666.65	$225,732.78	62.94%
163	$1,798.65	$1,128.66	$669.99	$225,062.79	62.75%
164	$1,798.65	$1,125.31	$673.34	$224,389.45	62.56%
165	$1,798.65	$1,121.95	$676.70	$223,712.75	62.38%
166	$1,798.65	$1,118.56	$680.09	$223,032.66	62.19%
167	$1,798.65	$1,115.16	$683.49	$222,349.17	62.00%

168	$1,798.65	$1,111.75	$686.91	$221,662.27	61.81%	14 years
169	$1,798.65	$1,108.31	$690.34	$220,971.93	61.62%	
170	$1,798.65	$1,104.86	$693.79	$220,278.14	61.43%	
171	$1,798.65	$1,101.39	$697.26	$219,580.88	61.23%	
172	$1,798.65	$1,097.90	$700.75	$218,880.13	61.04%	
173	$1,798.65	$1,094.40	$704.25	$218,175.88	60.85%	
174	$1,798.65	$1,090.88	$707.77	$217,468.10	60.65%	
175	$1,798.65	$1,087.34	$711.31	$216,756.79	60.45%	
176	$1,798.65	$1,083.78	$714.87	$216,041.93	60.26%	
177	$1,798.65	$1,080.21	$718.44	$215,323.48	60.06%	
178	$1,798.65	$1,076.62	$722.03	$214,601.45	59.86%	
179	$1,798.65	$1,073.01	$725.64	$213,875.81	59.66%	
180	$1,798.65	$1,069.38	$729.27	$213,146.53	59.45%	15 years
181	$1,798.65	$1,065.73	$732.92	$212,413.61	59.25%	
182	$1,798.65	$1,062.07	$736.58	$211,677.03	59.05%	
183	$1,798.65	$1,058.39	$740.27	$210,936.76	58.84%	
184	$1,798.65	$1,054.68	$743.97	$210,192.80	58.64%	
185	$1,798.65	$1,050.96	$747.69	$209,445.11	58.43%	
186	$1,798.65	$1,047.23	$751.43	$208,693.68	58.22%	
187	$1,798.65	$1,043.47	$755.18	$207,938.50	58.01%	
188	$1,798.65	$1,039.69	$758.96	$207,179.54	57.80%	
189	$1,798.65	$1,035.90	$762.75	$206,416.79	57.59%	
190	$1,798.65	$1,032.08	$766.57	$205,650.22	57.38%	
191	$1,798.65	$1,028.25	$770.40	$204,879.82	57.17%	
192	$1,798.65	$1,024.40	$774.25	$204,105.57	56.95%	16 years
193	$1,798.65	$1,020.53	$778.12	$203,327.44	56.74%	
194	$1,798.65	$1,016.64	$782.01	$202,545.43	56.52%	
195	$1,798.65	$1,012.73	$785.92	$201,759.50	56.30%	
196	$1,798.65	$1,008.80	$789.85	$200,969.65	56.09%	
197	$1,798.65	$1,004.85	$793.80	$200,175.85	55.87%	
198	$1,798.65	$1,000.88	$797.77	$199,378.07	55.65%	
199	$1,798.65	$996.89	$801.76	$198,576.31	55.42%	
200	$1,798.65	$992.88	$805.77	$197,770.54	55.20%	
201	$1,798.65	$988.85	$809.80	$196,960.74	54.98%	
202	$1,798.65	$984.80	$813.85	$196,146.90	54.75%	
203	$1,798.65	$980.73	$817.92	$195,328.98	54.53%	
204	$1,798.65	$976.64	$822.01	$194,506.97	54.30%	17 years
205	$1,798.65	$972.53	$826.12	$193,680.86	54.07%	
206	$1,798.65	$968.40	$830.25	$192,850.61	53.84%	
207	$1,798.65	$964.25	$834.40	$192,016.21	53.61%	
208	$1,798.65	$960.08	$838.57	$191,177.64	53.38%	
209	$1,798.65	$955.89	$842.76	$190,334.88	53.14%	
210	$1,798.65	$951.67	$846.98	$189,487.90	52.91%	
211	$1,798.65	$947.44	$851.21	$188,636.69	52.67%	
212	$1,798.65	$943.18	$855.47	$187,781.22	52.44%	
213	$1,798.65	$938.91	$859.75	$186,921.47	52.20%	
214	$1,798.65	$934.61	$864.04	$186,057.43	51.96%	

215	$1,798.65	$930.29	$868.36	$185,189.06	51.72%
216	$1,798.65	$925.95	$872.71	$184,316.36	51.48% **18 years**
217	$1,798.65	$921.58	$877.07	$183,439.29	51.24%
218	$1,798.65	$917.20	$881.46	$182,557.83	50.99%
219	$1,798.65	$912.79	$885.86	$181,671.97	50.75%
220	$1,798.65	$908.36	$890.29	$180,781.68	50.50%
221	$1,798.65	$903.91	$894.74	$179,886.94	50.25%
222	$1,798.65	$899.43	$899.22	$178,987.72	50.01%
223	$1,798.65	$894.94	$903.71	$178,084.01	49.76%
224	$1,798.65	$890.42	$908.23	$177,175.77	49.50%
225	$1,798.65	$885.88	$912.77	$176,263.00	49.25%
226	$1,798.65	$881.32	$917.34	$175,345.67	49.00%
227	$1,798.65	$876.73	$921.92	$174,423.74	48.74%
228	$1,798.65	$872.12	$926.53	$173,497.21	48.49% **19 years**
229	$1,798.65	$867.49	$931.17	$172,566.04	48.23%
230	$1,798.65	$862.83	$935.82	$171,630.22	47.97%
231	$1,798.65	$858.15	$940.50	$170,689.72	47.71%
232	$1,798.65	$853.45	$945.20	$169,744.52	47.45%
233	$1,798.65	$848.72	$949.93	$168,794.59	47.19%
234	$1,798.65	$843.97	$954.68	$167,839.91	46.92%
235	$1,798.65	$839.20	$959.45	$166,880.46	46.66%
236	$1,798.65	$834.40	$964.25	$165,916.21	46.39%
237	$1,798.65	$829.58	$969.07	$164,947.14	46.12%
238	$1,798.65	$824.74	$973.92	$163,973.22	45.85%
239	$1,798.65	$819.87	$978.79	$162,994.44	45.58%
240	$1,798.65	$814.97	$983.68	$162,010.76	45.31% **20 years**
241	$1,798.65	$810.05	$988.60	$161,022.16	45.04%
242	$1,798.65	$805.11	$993.54	$160,028.62	44.76%
243	$1,798.65	$800.14	$998.51	$159,030.11	44.49%
244	$1,798.65	$795.15	$1,003.50	$158,026.61	44.21%
245	$1,798.65	$790.13	$1,008.52	$157,018.09	43.93%
246	$1,798.65	$785.09	$1,013.56	$156,004.53	43.65%
247	$1,798.65	$780.02	$1,018.63	$154,985.90	43.37%
248	$1,798.65	$774.93	$1,023.72	$153,962.18	43.08%
249	$1,798.65	$769.81	$1,028.84	$152,933.34	42.80%
250	$1,798.65	$764.67	$1,033.98	$151,899.35	42.51%
251	$1,798.65	$759.50	$1,039.15	$150,860.20	42.23%
252	$1,798.65	$754.30	$1,044.35	$149,815.85	41.94% **21 years**
253	$1,798.65	$749.08	$1,049.57	$148,766.28	41.65%
254	$1,798.65	$743.83	$1,054.82	$147,711.46	41.35%
255	$1,798.65	$738.56	$1,060.09	$146,651.36	41.06%
256	$1,798.65	$733.26	$1,065.39	$145,585.97	40.77%
257	$1,798.65	$727.93	$1,070.72	$144,515.25	40.47%
258	$1,798.65	$722.58	$1,076.08	$143,439.17	40.17%
259	$1,798.65	$717.20	$1,081.46	$142,357.71	39.87%
260	$1,798.65	$711.79	$1,086.86	$141,270.85	39.57%
261	$1,798.65	$706.35	$1,092.30	$140,178.55	39.27%

262	$1,798.65	$700.89	$1,097.76	$139,080.80	38.97%	
263	$1,798.65	$695.40	$1,103.25	$137,977.55	38.66%	
264	$1,798.65	$689.89	$1,108.76	$136,868.78	38.36%	22 years
265	$1,798.65	$684.34	$1,114.31	$135,754.48	38.05%	
266	$1,798.65	$678.77	$1,119.88	$134,634.60	37.74%	
267	$1,798.65	$673.17	$1,125.48	$133,509.12	37.43%	
268	$1,798.65	$667.55	$1,131.11	$132,378.01	37.11%	
269	$1,798.65	$661.89	$1,136.76	$131,241.25	36.80%	
270	$1,798.65	$656.21	$1,142.45	$130,098.81	36.48%	
271	$1,798.65	$650.49	$1,148.16	$128,950.65	36.17%	
272	$1,798.65	$644.75	$1,153.90	$127,796.75	35.85%	
273	$1,798.65	$638.98	$1,159.67	$126,637.08	35.53%	
274	$1,798.65	$633.19	$1,165.47	$125,471.62	35.20%	
275	$1,798.65	$627.36	$1,171.29	$124,300.32	34.88%	
276	$1,798.65	$621.50	$1,177.15	$123,123.17	34.55%	23 years
277	$1,798.65	$615.62	$1,183.04	$121,940.14	34.23%	
278	$1,798.65	$609.70	$1,188.95	$120,751.19	33.90%	
279	$1,798.65	$603.76	$1,194.90	$119,556.29	33.57%	
280	$1,798.65	$597.78	$1,200.87	$118,355.42	33.23%	
281	$1,798.65	$591.78	$1,206.87	$117,148.55	32.90%	
282	$1,798.65	$585.74	$1,212.91	$115,935.64	32.57%	
283	$1,798.65	$579.68	$1,218.97	$114,716.66	32.23%	
284	$1,798.65	$573.58	$1,225.07	$113,491.60	31.89%	
285	$1,798.65	$567.46	$1,231.19	$112,260.40	31.55%	
286	$1,798.65	$561.30	$1,237.35	$111,023.05	31.21%	
287	$1,798.65	$555.12	$1,243.54	$109,779.52	30.86%	
288	$1,798.65	$548.90	$1,249.75	$108,529.76	30.52%	24 years
289	$1,798.65	$542.65	$1,256.00	$107,273.76	30.17%	
290	$1,798.65	$536.37	$1,262.28	$106,011.48	29.82%	
291	$1,798.65	$530.06	$1,268.59	$104,742.88	29.47%	
292	$1,798.65	$523.71	$1,274.94	$103,467.94	29.12%	
293	$1,798.65	$517.34	$1,281.31	$102,186.63	28.76%	
294	$1,798.65	$510.93	$1,287.72	$100,898.91	28.41%	
295	$1,798.65	$504.49	$1,294.16	$99,604.76	28.05%	
296	$1,798.65	$498.02	$1,300.63	$98,304.13	27.69%	
297	$1,798.65	$491.52	$1,307.13	$96,997.00	27.33%	
298	$1,798.65	$484.98	$1,313.67	$95,683.33	26.96%	
299	$1,798.65	$478.42	$1,320.23	$94,363.10	26.60%	
300	$1,798.65	$471.82	$1,326.84	$93,036.26	26.23%	25 years
301	$1,798.65	$465.18	$1,333.47	$91,702.79	25.86%	
302	$1,798.65	$458.51	$1,340.14	$90,362.65	25.49%	
303	$1,798.65	$451.81	$1,346.84	$89,015.82	25.12%	
304	$1,798.65	$445.08	$1,353.57	$87,662.24	24.75%	
305	$1,798.65	$438.31	$1,360.34	$86,301.90	24.37%	
306	$1,798.65	$431.51	$1,367.14	$84,934.76	23.99%	
307	$1,798.65	$424.67	$1,373.98	$83,560.78	23.61%	
308	$1,798.65	$417.80	$1,380.85	$82,179.93	23.23%	

309	$1,798.65	$410.90	$1,387.75	$80,792.18	22.84%
310	$1,798.65	$403.96	$1,394.69	$79,397.49	22.46%
311	$1,798.65	$396.99	$1,401.66	$77,995.83	22.07%
312	$1,798.65	$389.98	$1,408.67	$76,587.16	21.68% **26 years**
313	$1,798.65	$382.94	$1,415.72	$75,171.44	21.29%
314	$1,798.65	$375.86	$1,422.79	$73,748.65	20.90%
315	$1,798.65	$368.74	$1,429.91	$72,318.74	20.50%
316	$1,798.65	$361.59	$1,437.06	$70,881.68	20.10%
317	$1,798.65	$354.41	$1,444.24	$69,437.44	19.70%
318	$1,798.65	$347.19	$1,451.46	$67,985.97	19.30%
319	$1,798.65	$339.93	$1,458.72	$66,527.25	18.90%
320	$1,798.65	$332.64	$1,466.02	$65,061.23	18.49%
321	$1,798.65	$325.31	$1,473.35	$63,587.89	18.09%
322	$1,798.65	$317.94	$1,480.71	$62,107.18	17.68%
323	$1,798.65	$310.54	$1,488.12	$60,619.06	17.26%
324	$1,798.65	$303.10	$1,495.56	$59,123.51	16.85% **27 years**
325	$1,798.65	$295.62	$1,503.03	$57,620.47	16.44%
326	$1,798.65	$288.10	$1,510.55	$56,109.92	16.02%
327	$1,798.65	$280.55	$1,518.10	$54,591.82	15.60%
328	$1,798.65	$272.96	$1,525.69	$53,066.13	15.18%
329	$1,798.65	$265.33	$1,533.32	$51,532.81	14.75%
330	$1,798.65	$257.66	$1,540.99	$49,991.82	14.33%
331	$1,798.65	$249.96	$1,548.69	$48,443.13	13.90%
332	$1,798.65	$242.22	$1,556.44	$46,886.69	13.47%
333	$1,798.65	$234.43	$1,564.22	$45,322.47	13.03%
334	$1,798.65	$226.61	$1,572.04	$43,750.43	12.60%
335	$1,798.65	$218.75	$1,579.90	$42,170.53	12.16%
336	$1,798.65	$210.85	$1,587.80	$40,582.73	11.72% **28 years**
337	$1,798.65	$202.91	$1,595.74	$38,987.00	11.28%
338	$1,798.65	$194.93	$1,603.72	$37,383.28	10.84%
339	$1,798.65	$186.92	$1,611.74	$35,771.55	10.39%
340	$1,798.65	$178.86	$1,619.79	$34,151.75	9.94%
341	$1,798.65	$170.76	$1,627.89	$32,523.86	9.49%
342	$1,798.65	$162.62	$1,636.03	$30,887.83	9.04%
343	$1,798.65	$154.44	$1,644.21	$29,243.61	8.59%
344	$1,798.65	$146.22	$1,652.43	$27,591.18	8.13%
345	$1,798.65	$137.96	$1,660.70	$25,930.48	7.67%
346	$1,798.65	$129.65	$1,669.00	$24,261.49	7.21%
347	$1,798.65	$121.31	$1,677.34	$22,584.14	6.74%
348	$1,798.65	$112.92	$1,685.73	$20,898.41	6.28% **29 years**
349	$1,798.65	$104.49	$1,694.16	$19,204.25	5.81%
350	$1,798.65	$96.02	$1,702.63	$17,501.62	5.34%
351	$1,798.65	$87.51	$1,711.14	$15,790.48	4.87%
352	$1,798.65	$78.95	$1,719.70	$14,070.78	4.39%
353	$1,798.65	$70.35	$1,728.30	$12,342.48	3.91%
354	$1,798.65	$61.71	$1,736.94	$10,605.54	3.43%
355	$1,798.65	$53.03	$1,745.62	$8,859.92	2.95%

356	$1,798.65	$44.30	$1,754.35	$7,105.57	2.46%	
357	$1,798.65	$35.53	$1,763.12	$5,342.44	1.98%	
358	$1,798.65	$26.71	$1,771.94	$3,570.50	1.49%	
359	$1,798.65	$17.85	$1,780.80	$1,789.70	0.99%	
360	$1,798.65	$8.95	$1,789.70	$0.00	0.50%	30 years

TOTALS	$647,514.57	$347,514.57	$300,000.00	<---THESE ARE	30 year totals

Complete Credit Card Payment Schedule (Session #24)

Spreadsheet by BRENDA and GLEN

Credit Card Example

Amount owed	$1,400.00
Yearly rate(nominal)	0.24
Monthly rate	0.02
Minimum monthly payment	$30.00

Month	Payment	Interest	Princ. Repaid	Still owe
0				$1,400.00
1	$30.00	$0.00	$30.00	$1,370.00
2	$30.00	$27.40	$2.60	$1,367.40
3	$30.00	$27.35	$2.65	$1,364.75
4	$30.00	$27.29	$2.71	$1,362.04
5	$30.00	$27.24	$2.76	$1,359.28
6	$30.00	$27.19	$2.81	$1,356.47
7	$30.00	$27.13	$2.87	$1,353.60
8	$30.00	$27.07	$2.93	$1,350.67
9	$30.00	$27.01	$2.99	$1,347.68
10	$30.00	$26.95	$3.05	$1,344.64
11	$30.00	$26.89	$3.11	$1,341.53
12	$30.00	$26.83	$3.17	$1,338.36
13	$30.00	$26.77	$3.23	$1,335.13
14	$30.00	$26.70	$3.30	$1,331.83
15	$30.00	$26.64	$3.36	$1,328.47
16	$30.00	$26.57	$3.43	$1,325.04
17	$30.00	$26.50	$3.50	$1,321.54
18	$30.00	$26.43	$3.57	$1,317.97
19	$30.00	$26.36	$3.64	$1,314.33
20	$30.00	$26.29	$3.71	$1,310.61
21	$30.00	$26.21	$3.79	$1,306.83
22	$30.00	$26.14	$3.86	$1,302.96
23	$30.00	$26.06	$3.94	$1,299.02
24	$30.00	$25.98	$4.02	$1,295.00
25	$30.00	$25.90	$4.10	$1,290.90
26	$30.00	$25.82	$4.18	$1,286.72
27	$30.00	$25.73	$4.27	$1,282.46
28	$30.00	$25.65	$4.35	$1,278.10
29	$30.00	$25.56	$4.44	$1,273.67
30	$30.00	$25.47	$4.53	$1,269.14
31	$30.00	$25.38	$4.62	$1,264.52
32	$30.00	$25.29	$4.71	$1,259.81

33	$30.00	$25.20	$4.80	$1,255.01
34	$30.00	$25.10	$4.90	$1,250.11
35	$30.00	$25.00	$5.00	$1,245.11
36	$30.00	$24.90	$5.10	$1,240.01
37	$30.00	$24.80	$5.20	$1,234.81
38	$30.00	$24.70	$5.30	$1,229.51
39	$30.00	$24.59	$5.41	$1,224.10
40	$30.00	$24.48	$5.52	$1,218.58
41	$30.00	$24.37	$5.63	$1,212.95
42	$30.00	$24.26	$5.74	$1,207.21
43	$30.00	$24.14	$5.86	$1,201.36
44	$30.00	$24.03	$5.97	$1,195.39
45	$30.00	$23.91	$6.09	$1,189.29
46	$30.00	$23.79	$6.21	$1,183.08
47	$30.00	$23.66	$6.34	$1,176.74
48	$30.00	$23.53	$6.47	$1,170.28
49	$30.00	$23.41	$6.59	$1,163.68
50	$30.00	$23.27	$6.73	$1,156.95
51	$30.00	$23.14	$6.86	$1,150.09
52	$30.00	$23.00	$7.00	$1,143.10
53	$30.00	$22.86	$7.14	$1,135.96
54	$30.00	$22.72	$7.28	$1,128.68
55	$30.00	$22.57	$7.43	$1,121.25
56	$30.00	$22.43	$7.57	$1,113.68
57	$30.00	$22.27	$7.73	$1,105.95
58	$30.00	$22.12	$7.88	$1,098.07
59	$30.00	$21.96	$8.04	$1,090.03
60	$30.00	$21.80	$8.20	$1,081.83
61	$30.00	$21.64	$8.36	$1,073.47
62	$30.00	$21.47	$8.53	$1,064.94
63	$30.00	$21.30	$8.70	$1,056.23
64	$30.00	$21.12	$8.88	$1,047.36
65	$30.00	$20.95	$9.05	$1,038.31
66	$30.00	$20.77	$9.23	$1,029.07
67	$30.00	$20.58	$9.42	$1,019.65
68	$30.00	$20.39	$9.61	$1,010.05
69	$30.00	$20.20	$9.80	$1,000.25
70	$30.00	$20.00	$10.00	$990.25
71	$30.00	$19.81	$10.19	$980.06
72	$30.00	$19.60	$10.40	$969.66
73	$30.00	$19.39	$10.61	$959.05
74	$30.00	$19.18	$10.82	$948.23
75	$30.00	$18.96	$11.04	$937.20
76	$30.00	$18.74	$11.26	$925.94
77	$30.00	$18.52	$11.48	$914.46
78	$30.00	$18.29	$11.71	$902.75
79	$30.00	$18.05	$11.95	$890.80
80	$30.00	$17.82	$12.18	$878.62
81	$30.00	$17.57	$12.43	$866.19
82	$30.00	$17.32	$12.68	$853.52
83	$30.00	$17.07	$12.93	$840.59
84	$30.00	$16.81	$13.19	$827.40
85	$30.00	$16.55	$13.45	$813.95

86	$30.00	$16.28	$13.72	$800.23
87	$30.00	$16.00	$14.00	$786.23
88	$30.00	$15.72	$14.28	$771.95
89	$30.00	$15.44	$14.56	$757.39
90	$30.00	$15.15	$14.85	$742.54
91	$30.00	$14.85	$15.15	$727.39
92	$30.00	$14.55	$15.45	$711.94
93	$30.00	$14.24	$15.76	$696.18
94	$30.00	$13.92	$16.08	$680.10
95	$30.00	$13.60	$16.40	$663.71
96	$30.00	$13.27	$16.73	$646.98
97	$30.00	$12.94	$17.06	$629.92
98	$30.00	$12.60	$17.40	$612.52
99	$30.00	$12.25	$17.75	$594.77
100	$30.00	$11.90	$18.10	$576.66
101	$30.00	$11.53	$18.47	$558.20
102	$30.00	$11.16	$18.84	$539.36
103	$30.00	$10.79	$19.21	$520.15
104	$30.00	$10.40	$19.60	$500.55
105	$30.00	$10.01	$19.99	$480.56
106	$30.00	$9.61	$20.39	$460.17
107	$30.00	$9.20	$20.80	$439.38
108	$30.00	$8.79	$21.21	$418.16
109	$30.00	$8.36	$21.64	$396.53
110	$30.00	$7.93	$22.07	$374.46
111	$30.00	$7.49	$22.51	$351.95
112	$30.00	$7.04	$22.96	$328.99
113	$30.00	$6.58	$23.42	$305.56
114	$30.00	$6.11	$23.89	$281.68
115	$30.00	$5.63	$24.37	$257.31
116	$30.00	$5.15	$24.85	$232.46
117	$30.00	$4.65	$25.35	$207.10
118	$30.00	$4.14	$25.86	$181.25
119	$30.00	$3.62	$26.38	$154.87
120	$30.00	$3.10	$26.90	$127.97
121	$30.00	$2.56	$27.44	$100.53
122	$30.00	$2.01	$27.99	$72.54
123	$30.00	$1.45	$28.55	$43.99
124	$30.00	$0.88	$29.12	$14.87
125	$15.17	$0.30	$14.87	$0.00
Totals	$3,735.17	$2,335.17	$1,400.00	

APPENDIX C
Solutions and Comments for Selected Activities

Session #1 Activity Set

1. (a) 25; (b) 30; (c) 105; (d) 225; (e) 18; (f) 36; (g) 40; (h) 10; (i) 25; (j) 100.

Session #2 Activity Set

1. (a) 51; (b) 2; (c) 576; (d) 576; (e) 25; (f) 49.

Session #3 Activity Set

1. 4427.43264
3. Various possible responses.
4. (a) $1000 + (20)(0.05)($1000) = $2000; (b) $1000 + (20)(0.10)($1000) = $3000.
5. (a) $1000(1.05)^{20} = $2653.30; (b) $1000(1.10)^{20} = $6727.50.
6. (a) TRUE; (b) FALSE. See problems 4 and 5 for examples.
7. Years to double: 1%(70), 2%(35), 3%(24), 4%(18), 5%(15), 6%(12), 7%(11), 8%(9), 9%(9), 10%(8).
8. $5000(1.06)^4(1.08)^6 = $10,016.96.

Session #4 Activity Set

1. (a) 9937.733255; (b) 5428.836182.
3. (a) $10,000/(1.08)^5 = $6805.83; (b) $80,000/(1.065)^{10} = $42,618.08; (c) $1,000,000/(1.058)^{20} = $323,807.31.
4. (a) $\log(10)/\log(1.06) = 39.52$ (40 years); $\log(4)/\log(1.049) = 28.98$ (29 years); (c) $\log(15000/3500)/\log(1.053) = 28.18$ (29 years).
5.

Rate	Deposit needed	Years required
1%	$671,653	695
2%	$452,480	349
3%	$306,557	234
4%	$208,289	177
5%	$142,046	142
6%	$97,222	119
7%	$66,780	103
8%	$46,031	90
9%	$31,838	81
10%	$22,095	73
11%	$15,384	67
12%	$10,747	61
13%	$7,531	57
14%	$5,924	53
15%	$3,733	50
16%	$2,640	47
17%	$1,873	44
18%	$1,333	42
19%	$951	40
20%	$680	38

Session #5 Activity Set

1. $3^{(1/12)}-1 = .0958726911$; $3^1/12-1 = -.75$. The order of operations has $3^1/12-1$ evaluated as $3/12 - 1 = 0.25 - 1 = -0.75$.
3. Algebraically, $x = \$50,000/1.075^{20} = \$11,771$ (to nearest dollar).
4. Algebraically, $x = \log(5)/\log(1.084) = 19.9539$ (20 years).
5. Algebraically, $x = 400^{1/30} - 1 = 0.2210553$, or about 22.11%.
6. $2500(1.054)^x = 50,000 ==> x = \log(20)/\log(1.054) = 56.96$ (57 years).
7. $2000(1+x)^{10} = 16,800 ==> x = (16,800/2000)^{1/10} - 1 = 0.2372$, or about 23.72%.
8. $x(1.063)^8 = \$25,000 ==> x = \$15,334.65$.

HERKIMERS QUICK QUIZ (SESSIONS 1-5)

1. D3   2. E5   3. E4   4. E1   5. C4   6. C2   7. A1   8. B1   9. B5   10. D2

Session #6 Activity Set

1. (a) 5.141661255; (b) .4423009644
3. (a) $\$100(4.71712) = \$471.71$; (b) $\$100,000(2.40985) = \$240,985$;
   (c) 11 years; (d) 15 years.
4. (a) $\$10,000(0.29925) = \$2992.50$; (b) $\$1,000,000(0.21455) = \$214,550$.
5. (a) $(1.11)^{20} = 8.06231$; (b) $(1.11)^{-20} = 0.12403$; (c) The numbers in (a) and (b) are reciprocals.
6. (a) $\$10,000(2.25219) - \$10,000 = \$12,521.90$;
   (b) $\$100,000(4.66096) - 100,000 = \$366,096$.

Session #7 Activity Set

1. (a) 2.158924997; (b) .6499313863.

3.

$1.12	$1.20	$1.25	$1.30	$1.50
$1.76	$2.49	$3.05	$3.71	$7.59
$3.11	$6.19	$9.31	$13.79	$57.67
$9.65	$38.34	$86.74	$190.05	$3,325.26
$17.00	$95.40	$264.70	$705.64	$25,251.17

4.

$0.89	$0.83	$0.80	$0.77	$0.67
$0.56	$0.40	$0.33	$0.27	$0.13
$0.32	$0.16	$0.11	0.07	$0.02
$0.10	$0.03	$0.01	$0.01	$0.00
$0.06	$0.01	$0.00	$0.00	$0.00

Session #8 Activity Set

1. (a) 69858.21319; (b) 98272.91639.
3. (a) $25,000(1.09)[(1.09)^{15} - 1]/.09 = \$800,085$.
   (b) $10,000(1.072)[(1.072)^{30} - 1]/.072 = \$1,049,798$.
   (c) $1000(1.086)[(1.086)^{50} - 1]/.086 = \$768,681$
4. (a) $20,000(1.0735)^{20} = \$82,617$. Interest = $62,617.
   (b) $1000(1.0735)[(1.0735)^{20} - 1]/.0735 = \$45,728$. Interest = 25,728.
   (c) In (b), the entire $20,000 wasn't earning interest for the 20-year period.
5. (a) $4000(1.06)[(1.06)^{10} - 1]/.06 = \$55,887$. Interest = $15,887.
   (b) $4000(1.12)[(1.12)^{10} - 1]/.12 = \$78,618$. Interest = $38,618.
   (c) No. Interest in (b) is more than double the interest in (a).
6. Investment X accumulates to $1000(1.10)[(1.10)^{10} - 1]/.10 = \$17,531.17$.
   Interest earned = $7,531.17.
   Investment Y accumulates to $2000(1.05)[(1.05)^{10} - 1]/.05 = \$26,413.57$.
   Interest earned = $6,413.57.
   (a) Investment Y; (b) Investment X.

Session #9 Activity Set

1. 1578.300505.
3. $859.75. Solve linear equation $x(1.06)[(1.06)^{25} - 1]/.06 = 50,000$.
4. $4437.43. Solve linear equation $x(1.072)[(1.072)^{40} - 1]/.072 = 1,000,000$.
5. 10.04%. Solve $3000(1+i)[(1+i)^{20} - 1]/i = 190,000$ by trial and error method.
6. 5.00%. Solve $1000(1+i)[(1+i)^{50} - 1]/i = 220,000$ by trial and error method.

Session #10 Activity Set

1. 3827.010546.
3. $800[(1.11)^{50} - 1]/.11 = \$1,335,016.92$. Interest = $1,295,016.92.
4. $10,000(1.076)[(1.076)^{25} - 1]/.076 + \$10,000= \$752,122.29$. Interest = $492,122.29.
5. $1000[(1.065)^{20} - 1]/.065 = \$38,825.31$. Interest = $18,825.31.
6. $600(1.058)[(1.058)^{40} - 1]/.058 = \$93,439.58$. Interest = $69,439.58.
7. $5000(1.127)[(1.127)^{15} - 1]/.127 + \$5000= \$227,284.44$. Interest = $147,284.44.

HERKIMERS QUICK QUIZ (SESSIONS 6-10)
1. E5   2. D3   3. A2   4. B5   5. E2   6. D5   7. E3   8. B4   9. A4   10. B1

Session #11 Activity Set

1. 19601.27726
3. (a) $500(1.07)^{12} + \$700(1.07)^9 + \$900(1.07)^7 + \$200(1.07)^5 + \$600(1.07) = \$4,780.73$.
   (b) $10,000(1.0585)^8 + \$20,000(1.585)^6 + \$50,000(1.585)^4 + \$60,000(1.585)^2 =$
       $173,881.92.
   (c) $50,000(1.067)^5 + \$40,000(1.067)^4 + \$30,000(1.067)^3 = \$157,439.31$.

4. (a) Solve $\$600(1+i)^{12} + \$700(1+i)^9 + \$200(1+i)^7 + \$800(1+i)^5 = \$30,000$.
   $i = 32.04\%$.
   (b) Solve $\$900(1+i)^9 + \$300(1+i)^8 + \$400(1+i)^6 + \$700(1+i)^4 + \$200(1+i)^2 = \$4000$.
   $i = 7.33\%$.

Session #12 Activity Set

1. (a) 3680.043526; (b) 3355.040699.
3. $\$10,000[1 - (1.072)^{-20}]/.072 = \$104,313.12$. Total interest earned = \$95,686.88.
4. $\$5000[1 - (1.049)^{-30}]/.049 = \$77,746.24$. Total interest earned = \$72,253.76.
5. (a) $\$20,000 = D[1 - (1.08)^{-10}]/.08 ==> D = \$2980.59$.
   (b) $\$20,000 = D[1 - (1.08)^{-20}]/.08 ==> D = \$2037.04$.
   (c) $\$20,000 = D[1 - (1.08)^{-30}]/.08 ==> D = \$1776.55$.
   (d) $\$20,000 = D[1 - (1.08)^{-50}]/.08 ==> D = \$1634.86$.
6. $\$1000/1.088 + \$2000/(1.088)^2 + \$5000/1.088)^3 + \$10,000/(1.088)^4 = \$13,627.41$.

Session #13 Activity Set

1. (a) 3992.710037; (b) 5130.350526.
3. Solve $1400[1-(1+i)^{-25}]/i = 20,000$. Solution i= 4.87%.
4. Solve $1700[1-(1+i)^{-10}]/i = 10,000$. Solution i= 11.03%.
5. Solve $40,000/(1+i)^5 + 80,000/(1+i)^{10} = 50,000$. Solution i = 11.54%.
6. As the lender, you would prefer Option #1 since you get your money back quicker. The respective annual yields for the lender are 24.10%, 19.17%, and 13.62%. These are solutions to the following equations:
   Option #1: $1200 = 500/(1+i) + 500/(1+i)^2 + 500/(1+i)^3 + 500/(1+i)^4$
   Option #2: $1200 = 1000/(1+i)^2 + 1000/(1+i)^4$.
   Option #3: $1200 = 2000/(1+i)^4$.

Session #14 Activity Set

1. (a) 8.51356372; (b) 57.27499949
3. (a) $\$10,000(8.85) = \$88,500$. Interest = $(16)(\$10,000) - \$88,500 = \$71,500$.
   (b) $\$1000(10.59) = \$10,590$. Interest = $(20)(\$1000) - \$10,590 = \$9,410$.
   (c) $\$100,000(7.36) = \$736,000$. Interest = $(10)(\$100,000) - \$736,000 = \$264,000$.
4. (a) $\$10,000(37.45) = \$374,500$. Interest = $\$374,500 - 18(\$10,000) = \$194,500$.
   (b) $\$100,000(41.00) = \$4,100,000$. Interest = $\$4,100,000 - 20(\$100,000) = \$2,100,000$.
   (c) $\$1000(23.28) = \$23,280$. Interest = $\$23,280 - 15(\$1000) = \$8,280$.
5. $(1+i)^n P = (1+i)^n[1 - (1+i)^{-n}]/i = [(1+i)^n - 1]/i = A$.

Session #15 Activity Set

1. (a) 12233.48464; (b) 573770.1564; (c) $12233.48464(1.08)^{50} = 573770.1564$.
3. (a) **P** = \$5879.83, **A** = \$10,408.76, $\$5879.83(1.074)^8 = \$10,408.76$.
   (b) **P** = \$6879.83, **A** = \$12,179.01, $\$6879.83(1.074)^8 = \$12,179.01$.

(c) $P = \$6314.94$, $A = \$11,179.01$, $\$6314.94(1.074)^8 = \$11,179.01$.

4. (a) $P = \$10,000/(1.069)^2 + \$20,000/(1.069)^4 = \$24,065.81$;
   $A = \$10,000(1.069)^4 + \$20,000/(1.069)^2 = \$35,914.25$.
   $\$24,065.81(1.069)^6 = \$35,914.25$.

(b) $P = \$1000/(1.069) + \$2000/(1.069)^2 + \$5000/(1.069)^5 = \$6267.24$;
   $A = \$1000(1.069)^5 + \$2000(1.069)^4 + \$5000(1.069) = \$9352.82$;
   $\$6267.24(1.069)^6 = \$9352.82$.

HERKIMERS QUICK QUIZ (SESSIONS 11-15)
1. C4   2. B4   3. A3   4. C5   5. E5   6. B5   7. C3   8. E2   9. A1   10. B1

Session #16 Activity Set

NOTE: Differences of $0.01 (or 1 cent) in some table values are due to the rounding process.

1. (a) 12854.62285; (b) 5477.32375.

3.

Yr	Payment	Interest	Principal	Princip. OS	% Int.
0				$5000.00	
1	$2081.74	$600.00	$1481.74	$3518.26	29%
2	$2081.74	$422.19	$1659.55	$1856.70	20%
3	$2081.74	$223.04	$1858.70	$0.00	11%
TOTALS	$6245.23	$1245.23	$5000.00		

4.

Yr	Payment	Interest	Principal	Princip. OS	% Int.
0				$12,000.00	
1	$3005.48	$960.00	$2045.48	$9954.52	32%
2	$3005.48	$796.36	$2209.12	$7745.41	27%
3	$3005.48	$619.63	$2385.84	$5359.56	21%
4	$3005.48	$428.76	$2576.71	$2782.85	14%
5	$3005.48	$222.63	$2782.85	$0.00	7%
TOTALS	$15,027.39	$3,027.39	$12,000.00		

5.

Yr	Payment	Interest	Principal	Princip. OS	% Int.
0				$75,000.00	
1	$19,286.81	$10,500.00	$8,786.81	$66,213.19	54%
2	$19,286.81	$9,269.85	$10,016.67	$56,196.22	48%
3	$19,286.81	$7,867.47	$11,419.34	$44,776.88	41%
4	$19,286.81	$6,268,76	$13,018.05	$31,758.83	33%
5	$19,286.81	$4,446.24	$14,840.58	$16,918.26	23%
6	$19,286.81	2,368.56	$16,918.26	$0.00	12%
TOTALS	$115,720.87	$40,720.87	$75,000.00		

Session #17 Activity Set

1. (a) 56411.26702; (b) 126722.2702.
3. Payment = $(0.12)(\$50,000)/[1-(1.12)^{-10}] = \$8,849,21$.
   (a) Principal in payment #3 = $\$8,849.21(1.12)^{-8} = \$3,574.05$, interest = $\$5,275.16$, principal outstanding = $\$8,849.21[1-(1.12)^{-7}]/.12 = \$40,385.64$.
   (b) Principal in payment #9 = $\$8,849.21(1.12)^{-2} = \$7,054.54$, interest = $\$1,794.67$, principal outstanding = $\$8,849.21[1-(1.12)^{-1}]/.12 = \$7,901.08$.
4. Payment = $(0.085)(\$20,000,000)/[1-(1.085)^{-50}] = \$1,729,266.79$.
   In payment #3, principal = $\$34,453.60$, interest = $\$1,694,813.19$, principal outstanding after payment = $\$19,904,525.15$.
   In payment #12, principal = $\$71,796.32$, interest = $\$1,657,470.47$, principal outstanding after payment = $\$19,427,856.21$.
   In payment #37, principal = $\$551,881.28$, interest = $\$1,177,385.51$, principal outstanding after payment = $\$13,299,712.90$.
   In payment #48, principal = $\$1,353,856.97$, interest = $\$375,409.82$, principal outstanding after payment = $\$3,062,729.09$.

Session #18 Activity Set

1. .0495927721; (b) .1304003578.
3. The APR's are $(1+.04/12)^{12} - 1 = 4.074\%$, $(1+.0833/12)^{12} - 1 = 8.655\%$, $(1+.14/12)^{12} - 1 = 14.934\%$, and $(1+.22/12)^{12} - 1 = 24.359\%$.
4. $\$100,000(1.09)^{40} = \$3,140,942.01$, $\$100,000(1+.09/12)^{480} = \$3,610,990.20$.
5. (a) $\$10,000(1.0934)^{25} = \$93,213.05$.
   (b) $\$10,000(1 + .0934/12)^{300} = \$102,365.01$.
   (c) $\$600(1.1765)^{15} = \$6,871.16$.
   (d) $\$600(1 +.1765/12)^{180} = \$8,309.45$.
6. (a) $x(1.1325)^{20} = \$500,000 ==> x = \$41,515.07$;
   (b) $x(1+.1325/12)^{240} = \$500,000 ==> x = \$35,842.40$.
7. (a) $\$1000(1+1/1) = \$2000$; (b) $\$1000(1+1/12)^{12} = \$2613.04$.

Session #19 Activity Set

1. (a) 27175.37468; (b) .2213358583.
3. The APR's are $(1+.18/2)^{2} - 1 = 18.81\%$, $(1+.09/365)^{365} - 1 = 9.416\%$, $(1+.13/4)^{4} - 1 = 13.6475\%$, $(1+.08/12)^{12} - 1 = 8.2999\%$, and $(1+.15/1000)^{1000} - 1 = 16.182\%$.
4. (a) APR = $(1+.10/365)^{365} - 1 = 10.5155\%$; (b) APR = $(1+.101/12)^{12} - 1 = 10.5809\%$; (c) APR = $(1+.1012/4)^{4} - 1 = 10.5105\%$. (b) is best and (c) is worst.
5. (a) $\$100,000(1+.09/2)^{80} = \$3,383,009$; (b) $\$100,000(1+.09/4)^{160} = \$3,516,662$; (c) $\$100,000(1+.09/12)^{480} = \$3,610,990$; (d) $\$100,000(1+.09/365)^{14600} = \$3,658,200$; (e) $\$100,000(1+.09/1000)^{40000} = \$3,659,231$.
6. (a) $\$500,000/(1+.08/2)^{80} = \$21,692$; (b) $\$500,000/(1+.08/4)^{160} = \$21,035$;

276

(c) $\$500,000/(1+.08/12)^{480} = \$20,599$; (d) $\$500,000/(1+.08/365)^{14600} = \$20,388$;
(e) $\$500,000/(1+.08/1000)^{40000} = \$20,384$.

7. (a) $\$1000(1 + 1/1) = \$2000$; (b) $\$1000(1 + 1/12)^{12} = \$2613$;
(c) $\$1000(1 + 1/365)^{365} = \$2715$; (d) $\$1000(1 + 1/1000)^{1000} = \$2717$.

Session #20 Activity Set

1. (a) .2682417946; (b) .0864878798.
3. (a) $i^{(4)} = 4[(1.15)^{1/4} - 1] = 14.2232\%$; (b) $i^{(12)} = 12[(1.15)^{1/12} - 1] = 14.0579\%$;
(c) $i^{(365)} = 365[(1.15)^{1/365} - 1] = 13.9788\%$.
4. (a) The true monthly rate is $24\%/12 = 2\%$.
   Accumulation $= \$2000(1.02)[(1.02)^{36}-1]/.02 = \$106,068.51$.
   (b) $i^{(12)} = 12[(1.24)^{1/12}-1] = 21.70509\%$ and the true monthly rate is
   $i^{(12)}/12 = 1.808758\%$.
   Accumulation $= \$2000(1.01808758)[(1.01808758)^{36}-1]/.01808758 = \$102,061.47$.
5. $i^{(12)} = 12[(1.096)^{1/12}-1] = 9.20182\%$ and the true monthly rate is $i^{(12)}/12 = 0.007668$.
   Accumulation
   $= \$10,000[(1.007668)^{56} + 1.007668)^{34} +1.007668)^{28} +1.007668)^{20} +1.007668)^{7}]$
   $= \$62,888.67$.
6. $(1 + .082/4)^{4} - 1 = 8.4556\%$.
7. $4[(1.0785)^{1/4} - 1] = 7.6289\%$, $12[(1.0785)^{1/12} - 1] = 7.5809\%$, $365[(1.0785)^{1/365} - 1] = 7.5579\%$.

HERKIMERS QUICK QUIZ (SESSIONS 16-20)
1. B1   2. D4   3. C3   4. D2   5. E3   6. E1   7. C4   8. C5   9. B4   10. D3

Session #21 Activity Set

1. (a) 508.7957226; (b) 355.911163.
3.

Loan Amount	$10,000	$20,000	$25,000	$30,000	$40,000
Loan Term (years)	2	3	4	5	5
Loan Term (months)	24	36	48	60	60
Loan rate $i^{(12)}$	8%	10%	13%	15%	18%
Monthly payment	$452.27	$645.34	$670.69	$713.70	$1,015.74
Total Payments	$10,854.48	$23,232.24	$32,193.12	$42,822.00	$60,944.40
Total Interest Paid	$854.48	$3,232.24	$7,193.12	$12,822.00	$20,944.40

4. The monthly payment is $42,000(.094/12)/[1 - (1+.094/12)^{-60}] =$ $880.03.

Pymt. #	Pymt.	Interest	Princ.	Princ. Outs.
6	$880.03	$307.08	$572.95	$38,628.41
34	$880.03	$167.17	$712.86	$20,628.38
55	$880.03	$40.25	$839.78	$4,298.59

5.

Month #	Payment	Interest	Principal	Princ. Outs.
0				$5,000.00
1	$451.29	$62.50	$388.79	$4,611.21
2	$451.29	$57.64	$393.65	$4,217.56
3	$451.29	$52.72	$398.57	$3,818.98
4	$451.29	$47.74	$403.55	$3,415.43
5	$451.29	$42.69	$408.60	$3,006.83
6	$451.29	$37.59	$413.71	$2,593.13
7	$451.29	$32.41	$418.88	$2,174.25
8	$451.29	$27.18	$424.11	$1,750.13
9	$451.29	$21.88	$429.41	$1,320.72
10	$451.29	$16.51	$434.78	$885.84
11	$451.29	$11.07	$440.22	$445.72
12	$451.29	$5.57	$445.72	$0.00
TOTALS	$5,415.50	$415.50	$5,000.00	

Session #22 Activity Set

1. (a) 3727.865678; (b) 3326.512476.

[Answers below are spreadsheet totals formatted to show only two decimal places. A reminder to those using calculators that answers you get may differ slightly from those provided if you use numbers rounded to two decimal places when computing totals.]

3. The money totals are displayed to the nearest dollar.\

Loan #	Loan Amount	Loan Term (Years)	Loan Rate	Monthly Payment	Total (All Payments)	Total Interest Paid
1	$800,000	30	6.40%	$5,004	$1,801,457	$1,001,457
2	$180,000	15	5.92%	$1,511	$272,011	$92,011
3	$600,000	20	6.18%	$4,361	$1,046,670	$446,670
4	$600,000	10	6.18%	$6,716	$805,871	$205,871
5	$350,000	20	4.80%	$2,271	$545,124	$195,124
6	$350,000	20	5.80%	$2,467	$592,150	$242,150
7	$350,000	20	6.80%	$2,672	$641,205	291.205
8	$470,000	30	7.15%	$3,174	$1,142,788	$672,788
9	$470,000	20	7.15%	$3,686	$884,722	$414,722
10	$470,000	15	7.15%	$4,264	$767,521	$297,521

4.

Loan Amount	Loan Term (Years)	Loan Rate	Monthly Payment	Total (All Payments)	Total Interest Paid
$300,000	30	6.2%	$1,837	$661,467	$361,467
$300,000	20	6.2%	$2,184	$524,172	$224,172

5.

Loan Amount	Loan Term (Years)	Loan Rate	Monthly Payment	Total (All Payments)	Total Interest Paid
$300,000	30	5.25%	$1,657	$596,380	$296,380
$300,000	30	6.25%	$1,847	$664,975	$364,975

Session #23 Activity Set

1. (a) 647514.5672; (b) 455682.6871.

4.

	15 year-mortgage	20-year mortgage	30-year mortgage
Monthly payment	$2,531.57	$2,149.29	$1,798.65
Total all payments	$455,682.69	$515,830.36	$647,514.57
Total interest paid	$155,682.69	$215,830.36	$347,514.57
Total interest in first 12 payments	$17,653.84	$17,782.12	$17,899.78
Total interest in middle 12 payments	$11,032.14	$12,065.82	$13,068.08
Total interest in final 12 payments	$964.70	$819.03	$685.41

5.

	5%	6%	7%	8%
Monthly payment	$2,147.29	$2,398.20	$2,661.21	$2,935.06
Total all payments	$773,024.40	$863,352.00	$958,035.60	$1,056,621.60
Total interest paid	$373,024.40	$463,352.00	$558,035.60	$656,621.60

6.

	5%	6%	7%	8%
Monthly payment	$3,163.17	$3,375.43	$3,595.31	$3,822.61
Total all payments	$569,370.60	$607,577.40	$647,156.36	$688,069.80
Total interest paid	$169,370.60	$207,577.40	$247,156.36	$288,069.80

. Session #24 Activity Set

1. 1367.4.

3.

Month	Payment	Interest Required	Principal Repaid	Still Owe
0				$5,000.00
1	$90.00	$0.00	$90.00	$4,910.00
2	$90.00	$81.83	$8.17	$4,901.83
3	$90.00	$81.70	$8.30	$4,893.53

4.

Month	Payment	Interest Required	Principal Repaid	Still Owe
0				$5,000.00
1	$200.00	$0.00	$200.00	$4,800.00
2	$600.00	$80.00	$520.00	$4,280.00
3	$100.00	$71.33	$28.67	$4,251.33

5.

Month	Payment	Interest Required	Principal Repaid	Still Owe
0				$5,000.00
1	$90.00	$0.00	$90.00	$4,910.00
2	$1,000.00	$81.83	$918.17	$3,991.83
3	$800.00	$66.53	$733.47	$3,258.36

6. To pay debt off at month #4, the payment would be $252.22+(0.1825/12)($252.22) = $256.06.

Month	Payment	Interest Required	Principal Repaid	Still Owe
0				$3,200.00
1	$1,000.00	$0.00	$1,000.00	$2,200.00
2	$1,000.00	$33.46	$966.54	$1,233.46
3	$1,000.00	$18.76	$981.24	$252.22
4	256.06	3.84	$252.22	$0.00

Session #25 Activity Set

1. 7939.0955.

5.

Month	Payment	Interest	Principal Repaid	Still Owe
0				$5,000.00
1	$75.00	$0.00	$75.00	$4,925.00
2	$75.00	$0.00	$75.00	$4,850.00
3	$75.00	$0.00	$75.00	$4,775.00
4	$75.00	$71.63	$3.37	$4,771.63
5	$75.00	$71.57	$3.43	$4,768.20
6	$75.00	$71.52	$3.48	$4,764.72

HERKIMERS QUICK QUIZ (SESSIONS 21-25)

1. D4   2. C2   3. B5   4. C4   5. E3   6. D3   7. B4   8. A3   9. E5   10. C5

Session #26 Activity Set

1. 492,900

3.

Year	Amount in fund	Amount withdrawn	Fund amount remaining	Fund amount with interest accumulation
1	$900,000	$50,000	$850,000	$896,750
2	$896,750	$50,000	$846,750	$893,321
3	$893,321	$50,000	$843,321	$889,704

*Fund will last 52 years.

4.

Year	Amount in fund	Amount withdrawn	Fund amount remaining	Fund amount with interest accumulation
1	$900,000	$70,000	$830,000	$875,650
2	$875,650	$70,000	$805,650	$849,961
3	$849,961	$70,000	$779,961	$822,859

*Fund will last for 20 years.

5.

Year	Amount in fund	Amount withdrawn	Fund amount remaining	Fund amount with interest accumulation
1	$1,200,000	$100,000	$1,100,000	$1,168,200
2	$1,168,200	$100,000	$1,068,200	$1,134,428
3	$1,134,428	$100,000	$1,034,428	$1,098,563

*Fund will last for 20 years.

6..

Year	Amount in fund	Amount withdrawn	Fund amount remaining	Fund amount with interest accumulation
1	$1,200,000	$100,000	$1,100,000	$1,157,200
2	$1,157,200	$100,000	$1,057,200	$1,112,174
3	$1,112,174	$100,000	$1,012,174	$1,064,807

*Fund will last for 17 years.

7.

Year	Amount in fund	Amount withdrawn	Fund amount remaining	Fund amount with interest accumulation
1	$800,000	$30,000	$770,000	$811,580
2	$811,580	$30,000	$781,580	$823,785
3	$823,785	$30,000	$793,785	$836,650

*Fund will increase indefinitely since withdrawals are less than annual interest earned.

Session #27 Activity Set

1. 398985.
5. If x is the annual rate compounded monthly, then the formula $300(1+x/12)[(1+x/12)^{480}-1]/(x/12)$ will provide the accumulation of the $300 monthly deposits for 40 years. (a) $355,770; (b) $459,714; (c) $600,435; (d) $792,037.
6. Subtracting $72,000 from the accumulated amounts in activity #3 yields $283,770, $387,714, $528,435, and $720,037.
7. $3,000(150) = $450,000; $3,000(162) = $486,000; $3,000(178) = $534,000; $3,000(200) = $600,000.

Session #28 Activity Set

1. (a) 11472.01741; (b) 8771.086579.
5. $50,000(1.08)^{-15} + $3,250[1-(1.08)^{-15}]/.08 = $43,580.

6. $25,000(1.058)^{-10} + $1,750[1-(1.058)^{-10}]/.058 = $27,229.
7. $1,000,000(1.07)^{-20} + $60,000[1-(1.07)^{-20}]/.07 = $894,060.
8. $75,000.

Session #29 Activity Set

1. (a) .0491150634; (b) .0307454965.
5. If i = yield, then i = $(B/A)^{1/N}$ - 1.
6. (a) If i = yield, $(1+i)^x = 2$ ==> $1+i = 2^{1/x}$ ==> $i = 2^{1/x}$ - 1.
   (b)

# Years	16	16.5	17	17.75	18	19	20	21	22
Yield	4.43%	4.29%	4.16%	3.98%	3.93%	3.72%	3.53%	3.36%	3.20%

Session #30 Activity Set

1. (a) 97491.20384; (b) 87295.71509.
3. $500,000(1+i)^{-10} + $20,000[1-(1+i)^{-10}]/i = $512,000; i = 3.71%.
4. $1,000,000(1+i)^{-25} + $50,000[1-(1+i)^{-25}]/i = $970,000; i = 5.22%.
5. $80,000(1+i)^{-5} + $4,800[1-(1+i)^{-5}]/i = $76,000; i = 7.23%.
6. $250,000(1+i)^{-20} + $17,500[1-(1+i)^{-20}]/i = $260,000; i = 6.63%.
7. $200,000(1+i)^{-5} + $13,000[1-(1+i)^{-5}]/i = $180,000; i = 9.08%.

HERKIMERS QUICK QUIZ (SESSIONS 26-30)
1. E3   2. B2   3. E5   4. A5   5. A1   6. D4   7. C5   8. B3   9. E4   10. D1

Session #31 Activity Set

1. (a) 100*1.5^22 = 748182.7643 and 100*1.5^23 = 1122274.146. Answer: 23 months.
   (b) 100*1.5^34 = 97073973.74 and 100*1.5^35 = 145610960.6. Answer: 35 months.

5. 12^9 = 5,159,780,352, so world population would be exceeded at level 10.

6.

Date	New Shares	Shares Outstanding	Added to Fund	Profits paid	Amount in Fund
1-Jan	20	20	$200,000		$200,000
31-Jan				$24,000	$176,000
1-Feb	40	60	$400,000		$576,000
28-Feb				$72,000	$504,000
1-Mar	80	140	$800,000		$1,304,000
31-Mar				$168,000	$1,136,000
1-Apr	160	300	$1,600,000		$2,736,000
30-Apr				$360,000	$2,376,000
1-May	320	620	$3,200,000		$5,576,000
31-May				$744,000	$4,832,000
1-Jun	640	1260	$6,400,000		$11,232,000
30-Jun				$1,512,000	$9,720,000
1-Jul	0	1260	$0		$9,720,000
31-Jul				$1,512,000	$8,208,000
1-Aug	0	1260	$0		$8,208,000
31-Aug				$1,512,000	$6,696,000
1-Sep	0	1260	$0		$6,696,000
30-Sep				$1,512,000	$5,184,000
1-Oct	0	1260	$0		$5,184,000
31-Oct				$1,512,000	$3,672,000
1-Nov	0	1260	$0		$3,672,000
30-Nov				$1,512,000	$2,160,000
1-Dec	0	1260	$0		$2,160,000
31-Dec				$1,512,000	$648,000

Session #32 Activity Set

1. (a) 4^3^2 is interpreted as (4^3)^2 = 4096 on many calculators;   (b) This attempted computation usually yields and OVERFLOW message.

Session #33 Activity Set

1. (a) 5.671417169; (b) 5.538461538.

4. $x = 0.0785$, or 7.85%. The corresponding time is 9.1756 years.

5.

x = annual interest rate	Actual Time Required (Yrs) Log(2)/Log(1+x)	Estimated time by *Rule of 72* 72/(100x)	% Error
0.25 (25%)	3.11	2.88	7.4%
0.30 (30%)	2.64	2.40	9.1%
0.35 (35%)	2.31	2.06	10.8%
0.40 (40%)	2.06	1.80	12.6%
0.50 (50%)	1.71	1.44	15.8%

6. Formulas are quite accurate in the 5% to 10% region. They become increasingly less accurate as interest rates increase beyond 20%.

1. (a) 10.91817654; (b) 113.47546.
5. (a) $i^{(6)}/6 = 20\%$ ==> True annual rate = $(1.2)^6 - 1 = 198.6\%$.
   (b) $i^{(3)}/3 = 30\%$ ==> True annual rate = $(1.3)^3 - 1 = 119.7\%$.
6.

Amount Borrowed	Loan Fee	True Rate for 2-Week Period	Nominal Annual Rate $i^{(26)}$	True Annual Rate
$100	$21	21%	546%	14,104%
$100	$22	22%	572%	17,494%
$100	$23	23%	598%	21,654%
$100	$24	24%	624%	26,751%
$100	$25	25%	650%	32,987%
$100	$26	26%	676%	40,604%
$100	$27	27%	702%	49,892%
$100	$28	28%	728%	61,200%
$100	$29	29%	754%	74,947%
$100	$30	30%	780%	91,633%

Session #35 Activity Set

1. (a) 575366.7237; (b) 552173.8994; (c) 1127540.623

3. (a) $391,824 (Scenario #1); $446,602 (Scenario #2); $838,426 (Scenario #3).
   (b) $842,574 (Scenario #1); $686,184 (Scenario #2); $1,528,759 (Scenario #3).
   (c) $2,603,855 (Scenario #1); $1,350,122 (Scenario #2); $3,953,977 (Scenario #3).
4. Plan 1: $10,000(1+i)[[(1+i)^{20}-1]/i](1+i)^{25}$. Plan 2: $300,000(1+i)^{25}$.
5. (a) $1,175,717 (Plan #1); $1,015,906 (Plan #2).
   (b) $2,380,753 (Plan #1); $1,628,230 (Plan #2).
   (c) $6,826,136 (Plan #1); $3,250,412 (Plan #2)

Session #36 Activity Set

1. (a) 208243.216; (b) 208327.7572; (c) 208328.7063
3. (a) $100,000(1.10)^{30} = \$1,744,940$.
   (b) $100,000(1 + 0.10/4)^{120} = \$1,935,815$.
   (c) $100,000(1 + 0.10/12)^{360} = \$1,983,740$.
   (d) $100,000(1 + 0.10/365)^{10950} = \$2,007,729$.
   (e) $100,000e^3 = \$2,008,554$.

4. (a) $\$1000(1.10)^{50} = \$117,391$; (b) $\$1000e^{(.10)(50)} = \$1000e^5 = \$148,413$.

5. (a) $(1.07)^x = 5$ ==> $x = \log(5)/\log(1.07) = 23.79$ years.
   (b) $e^{0.07x} = 5$ ==> $x = \ln(5)/0.07 = 22.99$ years. NOTE: $\ln(5) = \log_e(5)$.

6.

Year	Deposit	Accumulation at true rate of 10%	Accumulation at 10% compounded continuously
0	$10,000	$10,000	$10,000
1	$10,000	$21,000	$21,051
2	$10,000	$33,100	$33,266
3	$10,000	$46,410	$46,764
4	$10,000	$61,051	$61,683
5		$67,156	$68,170

Session #37 Activity Set

1. (a) 2796.858034; (b) 1679.34697

4. The monthly payment is $200,000(.072/12)/[1 - (1+.072/12)^{-180}] = $1,820.09. The total amount paid is $327,616.83.

5. Using the results from activity #4, the first, thirteenth, twenty-fifth, etc. payments would be $3,640.18. The loan would be paid off in 157 months, eliminating 23 months of payments. Total payments would be $307,874.96. The savings would be $19,741.87.

6. Using the results from activity #4, the 13th payment would be $11,820.09. This extra payment would have the loan paid off in 166 months. Total payments would be $311,531.78. The savings would be $16,085.05.

Session #38 Activity Set

1. (a) 297493.4051; (b) 294779.3133.

4.

	ORIGINAL MORTGAGE				
Loan Amount	$233,194.35	$233,000.00	$233,000.00	$233,000.00	$233,000.00
Loan Rate	7.5%	6%	6%	6%	6%
Term (Years)	24	30	25	20	15
Monthly Payment	$1,748.04	$1,396.95	$1,501.22	$1,669.28	$1,966.19
Total Payments	$503,435.52	$502,902.00	$450,366.00	$400,627.20	353,914.20
Closing costs	$0.00	$9,320.00	$9,320.00	$9,320.00	$9,320.00
Total Cost	$503,435.52	$512,222.00	$459,686.00	$409,947.20	$363,234.20

5.

	ORIGINAL MORTGAGE				
Loan Amount	$164,091.87	$164,000.00	$164,000.00	$164,000.00	$164,000.00
Loan Rate	7%	6%	5%	6%	5%
Term (Years)	20	20	20	15	15
Monthly Payment	$1,272.20	$1,174.95	$1,082.33	$1,383.93	$1,296.90
Total Payments	$305,328.00	$281,988.00	$259,759.20	$249,107.40	$233,442.00
Closing costs	$0.00	$8,200.00	$8,200.00	$8,200.00	$8,200.00
Total Cost	$305,328.00	$290,188.00	$267,959.20	$257,307.40	241,642.00

Session #39 Activity Set

1. (a) 53897.29057; (b) 83343.67507.
3. The semiannual interest required ($17,500) is equal to the coupon paid. One would pay $500,000 for this bond and there would be no write-up or write-down since the book value change would be $0 after each coupon payment.

4.

Coupon #	Year #	A = Interest Required	C = Coupon	Book Value Change (A-C)	Book Value
0	0				$189,848.62
1	0.5	$9,492.43	$8,000.00	$1,492.43	$191,341.05
2	1	$9,567.05	$8,000.00	$1,567.05	$192,908.10
3	1.5	$9,645.40	$8,000.00	$1,645.40	$194,553.50
4	2	$9,727.68	$8,000.00	$1,727.68	$196,281.18
5	2.5	$9,814.06	$8,000.00	$1,814.06	$198,095.24
6	3	$9,904.76	$8,000.00	$1,904.76	$200,000.00

5.

Coupon #	Year #	A = Interest Required	C = Coupon	Book Value Change (A-C)	Book Value
0	0				$216,524.39
1	0.5	$5,413.11	$8,000.00	-$2,586.89	$213,937.49
2	1	$5,348.44	$8,000.00	-$2,651.56	$211,285.92
3	1.5	$5,282.15	$8,000.00	-$2,717.85	$208,568.07
4	2	$5,214.20	$8,000.00	-$2,785.80	$205,782.27
5	2.5	$5,144.56	$8,000.00	-$2,855.44	$202,926.83
6	3	$5,073.17	$8,000.00	-$2,926.83	$200,000.00

Session #40 Activity Set

1. (a) 57011.31194; (b) 93478.045.
5. (a) 28 years, after which $18,252 would be in fund.
   (b) 20 years, last withdrawal = $58,272, remaining in fund $24,301.
   (c) 16 years, last withdrawal = $72,038, remaining in fund $56,339.

6. (a) 18 years, last withdrawal = $89,734, remaining in fund $72,631.
   (b) 23 years, last withdrawal = $122,489, remaining in fund $16,868.
   (c) 32 years, last withdrawal = $145,252, remaining in fund $86,924.

Session #41 Activity Set

1. (a) 35.2; (b) 194.493

5.

Month	Payment	Interest	Principal	Still Owe
0				$8,000.00
1	$150.00	$6.67	$143.33	$7,856.67
2	$150.00	$6.55	$143.45	$7,713.22
3	$150.00	$6.43	$143.57	$7,569.64
4	$150.00	$138.78	$11.22	$7,558.42
5	$150.00	$138.57	$11.43	$7,546.99
6	$150.00	$138.36	$11.64	$7,535.35

6. In the 7th payment, the requested amounts are $238.00, $62.00, and $10,138.00.
In the 8th payment, the requested amounts are $236.55, $63.45, and $10,074.55.

Session #42 Activity Set

1. (a) 2398.202101; (b) 2917.881734.

4.

Mortgage Amount	ARM Type	Mortgage Period (Years)	Initial Mortgage Rate	Initial Monthly Payment	Principal Outstanding After 1 Year	Adjusted Monthly Rate	New Monthly Payment
$240,000	1/1	30	5.2%	$1317.87	$236,584.98	6.8%	$1558.81

5.

Mortgage Amount	ARM Type	Mortgage Period (Years)	Initial Mortgage Rate	Initial Monthly Payment	Principal Outstanding After 5 Years	Adjusted Monthly Rate	New Monthly Payment
$320,000	5/1	30	6%	$1918.56	$297,773.94	7.6%	$2219.93

6.

Mortgage Amount	ARM Type	Mortgage Period (Years)	Initial Mortgage Rate	Initial Monthly Payment	Principal Outstanding After 5 Years	Adjusted Monthly Rate	New Monthly Payment
$190,000	5/5	20	5.3%	$1285.62	$159,405.12	6.75%	1410.59

7.

Mortgage Amount	ARM Type	Mortgage Period (Years)	Initial Mortgage Rate	Initial Monthly Payment	Principal Outstanding After 1 Year	Adjusted Monthly Rate	New Monthly Payment
$420,000	1/1	15	4.9%	$3299.50	$400,533.16	6.2%	3572.46

Herkimer's Final Financial Test

1. Payment = $\$200,000(.072)/[1 - (1.072)^{-25}] = \$17,472.44$. The interest portion of the first payment is $\$200,000(0.072) = \$14,400$.

2. (a) $\$340,000(0.0675/12)/[1 - (1 + 0.0675/12)^{-360}] = \$2,205.23$;
   (b) $\$328,355(0.0525/12)/[1 - (1 + 0.0525/12)^{-360}] = \$1,813.19$.

3. $\$15,000[(1.059)^{20} - 1]/.059 = \$545,888.83$.

4. $K[1 - (1+I)^{-T}]/I$.

5. 1776.

6. $\$10,000(1 + .0675/365)^{(365)(20)} = \$38,569.44$.

7. You would be saving 2^{30} quarters on the 31st day of the month. The amount would be $(\$0.25)(2^{30}) = \$286,435,456$.

8. $\$50,000(.085)/[1 - (1.085)^{-10}] = \$7,620.39$

9. 1935; basic purpose was to provide a source of money for the U.S. Treasury and a safe investment for people with modest incomes.

10. $1 + i = e^{0.0786} ==> i = e^{0.0786} - 1 = 8.177\%$.

11. $(\$600,000 - \$40,000)(1.058) = \$592,480$.

12. $\$8,000(1.065)^{20} = \$28,189.16$.

13. $\$2,400[1 - (1.07)^{-10}]/.07 + \$40,000(1.07)^{-10} = \$37,190.57$.

14. 1959.

15. Ten characters of which eight are numbers.

16. (a) $\$100,000(1.06)^{15} = \$239,655.82$; (b) $\$100,000(1 + .06/12)^{180} = \$245,409.36$;
    (c) $\$100,000e^{(15)(.06)} = \$100,000e^{0.9} = \$245,960.31$.

17. Using the $(1+i)^n$ table in Session #6, the answer is 11 years. One could also use a calculator and note that $(1.07)^{10} = 1.9672$ and $(1.07)^{11} = 2.105$.

18. $\$100,000(1+i)^{-10} + \$5,000[1 - (1 + i)^{-10}]/i = \$94,000$.

19. About 4:10, although some suggest it might be 2:21.

20. $100,000/(1.068)^{20} = \$26,827.18$.

21. The annual yield is $i = (1 + .0595/365)^{365} - 1 = 6.13\%$.

22. A Ponzi scheme is basically an investment swindle in which high profits are promised and early investors are initially paid profits with money raised from later investors or from a portion of the money they themselves invested.

23. Estimate is $72/[100(.0683)] = 10.54$ years. To find actual value x solve $(1.0683)^x = 2 \implies x = \log(2)/\log(1.0683) = 10.49$ years.

24. $\$20,000[1 - (1.064)^{-10}]/.064 = \$144,451.85$.

25. (a) $\$3,000[1-(1.04)^{-40}]/.04 + \$100,000(1.04)^{-40} = \$80,207.23$.
    (b) Interest required $= (\$80,207.23)(.04) = \$3,208.29$. The book value after the first coupon payment is $\$80,207.23 + (\$3,208.29 - \$3,000.00) = \$80,415.52$.

26. $(0.26/12)(\$2,940) = \$63.70$.

27. $\$10,000(1.0598)^{10} + \$20,000(1.0598)^8 + \$40,000(1.0598)^5 = \$103,182.14$.

28. $\$40,000, \$40,000(1.04) = \$41,600, \$40,000(1.04)^2 = \$43,264, \$40,000(1.04)^3 = \$44,995$.

29. $i^{(12)} = 12[(1.13)^{1/12} - 1] = 12.28\%$.

30. $(0.22/12)(\$623) = \$11.42$.

31. $\$60,000(.14/12)/[1 - (1 + .14/12)^{-120}] = \$931.60$.

32. $\$1000[1-(1.07)^{-12}]/.07 = \$7942.69$. (Using table in Session #14 containing decimals to two places would yield $7940.)

33. $i^{(26)}/26 = 80/500 = 0.16 = 16\%$. True annual rate $= (1.16)^{26} - 1 = 46.41412$, or 4641%.

34. The principal outstanding after the 3rd payment is $\$6,500 - 3(\$125) = \$6,125$. The interest portion in the 4th payment is $\$6,125(0.22/12) = \$112.29$.

35. $\$480,000(.0675/12)/[1 - (1+.0675/12)^{-360}] = \$3,113.27$.

36. $x(1.07)[(1.07)^{15}-1]/.07 = \$25,000 \implies x = \$929.78$.

37. The respective accumulations are $\$5,000(1.07)^{45} = \$105,012$ and $\$15,000(1.07)^{25} = \$81,411$.

38. (a) $\$320,000(.054/12)/[1-(1+.054/12)^{-360}] = \$1,796.90$.
    (b) $\$1,796.90[1-(1+.054/12)^{-348}]/(.054/12) = \$315,609.87$.
    (c) $\$315,609.87(.064/12)/[1-(1+.064/12)^{-348}] = \$1,996.91$.

39. $(12000/5000)^{1/7} - 1 = 13.32\%$.

40. $\$5000(1.06)[(1.06)^{20} - 1]/.06 = \$194,963.63$.

41. There are many possible responses. A primary advantage would be that since the extra payments reduce the outstanding principal the term of the mortgage is reduced. It will take less time to pay off the loan.

42. $\$100(1.036)^{10} = \$142.43$.

43. (a) $\$300,000(.057/12)/[1 - (1+.057/12)^{-360}] = \$1,741.20$.
    (b) $\$300,000(.067/12)/[1 - (1+.067/12)^{-360}] = \$1,935.83$.

44. $\$10,000(1.10)^{15} - \$10,000 = \$31,772.48$.

45. (a) $\$290,000(.049/12)/[1-(1+.049/12)^{-240}] = \$1,897.89$.
    (b) $\$1,897.89[1-(1+.049/12)^{-180}]/(.049/12) = \$241,586.55$.
    (c) $\$241,586.55(.062/12)/[1-(1+.062/12)^{-180}] = \$2,064.84$.

46. $\$50,000(1.06)^5(1.07)^5(1.08)^5 = \$137,891.34$.

47. $\$375(1 + i)^{6.25} = \$500 ==> i = (4/3)^{1/6.25} - 1 = 4.71\%$.

48. $(1.126)^x = 8 ==> x = \log(8)/\log(1.126) = 17.52$, or about 18 years.

49. $K[(1+I)^T - 1]/I$.

50. $\$50,000 = x[1 - (1.06)^{-20}]/.06 = \$4,359.23$.

51. $180(\$400,000)(.09/12)/[1 - (1+.09/12)^{-180}] - \$400,000 = \$330,272$.

52. Principal portion of payment = $\$15 - (0.25/12)(\$586) = \$15 - \$12.21 = \$2.79$. Principal outstanding after payment is $\$586 - \$2.79 = \$583.21$.

53. Using the $(1+i)^{-n}$ table in Session #6, the answer is $\$100,000(0.21455) = \$21,455$. One could also use a calculator and compute $\$100,000(1.08)^{-20}$.

54. After the first withdrawal the remaining amount of $\$760,000$ will accumulate to $\$760,000(1.06) = \$805,600$. The withdrawals of $\$40,000$ are more than covered by the interest paid.

55. $\$1000(1.07)[(1.07)^{40} - 1]/.07 = \$213,609.57$.

56. If the piece represents the larger portion of the bill (more than 50% of the original bill) it can legally be spent.

57. $80,000[1 - (1.06)^{-20}]/.06 + $1,000,000(1.06)^{-20} = $1,229,398.42$.

58. $240($500,000)(.069/12)/[1-(1+.069/12)^{-240}] = $923,169$.

59. (a) $3,500[1-(1.025)^{-30}]/.025 + $100,000(1.025)^{-30} = $120,930.29$.

   (b) Interest required = ($120,930.29)(.025) = $3,023.26$.. The book value after the first coupon payment is $120,930.29 + ($3,023.26- $3,500.00) = $120,453.55$.

60. (a) $300,000(.06/12)/[1 - (1+.06/12)^{-240}] = $2,149.29$.
    (b) $300,000(.06/12)/[1 - (1+.06/12)^{-360}] = $1,798.65$.

61. (a) At 7.2%, monthly payment = $420.000(.072/12)/[1 - (1 + .072/12)^{-360}] = $2,850.91$. The total of all payments is $360($2,850.91) = $1,026,327.60$.
    (b) At 5.2%, monthly payment = $420.000(.052/12)/[1 - (1 + .052/12)^{-360}] = $2,306.27$. The total of all payments is $360($2,306.27) = $830,257.20$.

62. $i^{(26)}/26 = 25/100 = 0.25 = 25\%$. True annual rate = $(1.25)^{26} - 1 = 329.8722$, or about 32,987%.

63. $100,000(1.10)^{10}(1+.10/12)^{60}(1+.10/365)^{3650} = $1,159,869.98$.

64. $1(1 + i)^{17} = $4 ==> i = 4^{1/17} - 1 = .08496$, or about 8.5%.

65. $24,000 = $1,500[1-(1+i)^{-10}]/i + $25,000(1+i)^{-10}$.

66. ($700,000 - $3,500)(1 + .06/12) = $699,982.50$.

67. $1,000[(1.07)^{20} - 1]/.07 = $40,995.49$. Using the table in Session #14 containing decimals to two places would yield $41,000$.)

68. Each payment = $20,000(0.15)/[1 - (1.15)^{-5}] = $5,966.31$. Total interest = $5($5,966.31) - $20,000 = $9,831.55$.

69. Principal outstanding after 6th payment is $8,100 - 6($200) = 6.900. Interest in 7th payment is $6,900(0.26/12) = 149.50. The principal portion of this payment is 50.50.

70. (a) $5000[[(1.06)^5 - 1]/.06](1.06)^2 = $31,669.19$;
    (b) $5000[[1-(1.06)^{-5}]/.06](1.06)^{-1} = $19,869.64$;
    (c) $19,869.64(1.06)^8 = $31,669.19$.

71. Payment = $200,000(.072)/[1 - (1.072)^{-25}] = $17,472.44. The interest portion of the first payment is $200,000(0.072) = $14,400. The principal repaid is $17,472.44 - $14,400.00 = 3,072.44. The principal outstanding after the first payment is $200,000 - $3,072.44 = 196,927.56.

72. Martha Washington.

73. $40.00 - (.23/12)($1960) = $2.43.

74. $20,000 = $30,000/(1+x)^6 + $50,000/(1+x)^{10}.

75. The outstanding balance will be reduced by $300. This will reduce the interest portion of all future payments and will decrease the term of the loan.

76. The annual yield is i = (1 + .12/12)^{12} - 1 = (1.01)^{12} - 1 = 12.68\%.

77. (a) $50,000(1.065)[(1.065)^5 - 1]/.065 = $303,186.38;
    (b) $50,000[e^{.065} + e^{.065(2)} + e^{.065(3)} + e^{.065(4)} + e^{.065(5)}] = $305,112.96.

78. $1,000,000(1 + i)^{-25} + $60,000[1 - (1 + i)^{-25}]/i = $1,150,000.

79. x(1 + .088/12)^{240} = $30,000 ==> x = $5,194.60.

80. $100(1.042)^{-12} = $61.04.

81. $1000(1+x)[(1+x)^{10}-1]/x = $21,000.

82. (a) $8600(.08/12) = $57.33; (b) $8600(.18/12) = $129.00;
    (c) $8600(.28/12) = $200. 67.

83. i^{(365)} = 365[(1.2)^{1/365} - 1] = 18.24\%.

84. $5000 + $5000(1.11)[(1.11)^5 - 1]/.11 = $39,564.30.

85. (a) $100,000(1.09)^{-20} = $17,843.09; (b) $100,000(1+.09/12)^{-240} = $16,641.28;
    (c) $100,000(1+.09/365)^{-7300} = $16,533.56.

86. A total of $25,000 is invested in investment #1 and accumulates to
    $5,000(1.07)[[(1.07)^5 - 1]/.07](1.07)^{40} = $460,711. A total of $150,000 is invested in
    investment #2 and accumulates to $10,000(1.07)[(1.07)^{15} - 1]/.07 = $268,881.
    Investment #1 has the greatest accumulation.

87. Alexander Hamilton and Benjamin Franklin.

88. For (A) the loan rate is $2^{1/5} - 1 = 14.87\%$. For (B) the loan rate is $4.5^{1/10} - 1 = 16.23\%$. In terms of yield rate, (A) is better for the borrower.

89. $\$6,000[1-(1.10)^{-5}]/.10 + \$100,000(1.10)^{-5} = \$84,836.85$.

90. (a) $\$250,000(.0625/12)/[1-(1+.0625/12)^{-360}] = 1539.29$.
    (b) $\$1539.29[1-(1+.0625/12)^{-359}]/(.0625/12) = \$249,762.30$.
    (c) $\$249,762.30(.0625/12) = \$1,300.85$.
    (d) $\$244,762.30(.0625/12) = \$1,274.80$.

91. $\$10,000(1+.062/365)^{-1825} + \$10,000(1+.062/365)^{-3650} = \$12,714.39$.

92. $i^{(12)}/12 = 50\%$. Hence annual yield $i = (1.5)^{12} - 1 = 12,875\%$.

93. $\$1000(1+i)^{19} = \$6000 \Longrightarrow i = 6^{1/19} - 1 = 9.89\%$.

94. (a) Payment $= \$470,000(.0575/12)/[1-(1+.0575/12)^{-360}] = \$2,742.79$; total payments $= \$987,404.40$.
    (b) Payment $= \$470,000(.0575/12)/[1-(1+.0575/12)^{-180}] = \$3,902.93$; total payments $= \$702,527.40$.

95. $\$8000 = \$4000(1+x)^{-3} + \$7000(1+x)^{-5}$.

96. The rate 8.12% compounded semiannually has an annual yield $= (1+.0812/2)^2 - 1 = 8.2848\%$; the rate 7.98% compounded daily has an annual yield $(1+.0798/365)^{365} - 1 = 8.3061\%$, which is the higher annual yield.

97. Thomas Jefferson.

98. $\$50,000/(1.025)^{15} = \$34,523$.

99. $\$250(1 + i)^{15.75} = 500 \Longrightarrow i = 2^{1/15.75} - 1 = 4.50\%$.

100. $\$1000(1.062)^{10} + \$2000(1.062)^{9} + \$3000(1.062)^{8} + \$4000(1.062)^{7} + \$5000(1.062)^{6} = \$23,383.57$.

INDEX OF TERMS (BY SESSION NUMBER)

Indicates that the concept is referenced or used in many of the following sessions

TOPIC	SESSIONS
accumulation of multiple deposits	8 *
adjustable rate mortgages	2, 32, 42
algebra	3, 4, 5, 8, 9, 10, 15, 16
Allen, Woody	36
American Express	26
amortization schedule	39
amortized value of a bond	39
Anderson, Thomas	41
annual rate compounded daily	19 *
annual rate compounded monthly	18 *
annual yield	13 *
Annuit Coeptis	13
APR	18 *
ARM	32, 42
bailout	1
Bank America	27
bankruptcy	1
bear market	1
bond amortization	32, 39
bond purchase price	28, 39
bond yields	29
book value of a bond	39
Borge, Victor	37
Brown, Becky	44
Bruce, Blanche	6
Bureau of Engraving and Printing	9
Carlin, George	21
certificate of deposit	18
Chase, Samuel P.	34
Chesterton, Gilbert	14
Civil War	3
Coinage Act of 1965	37
Colton, Charles Caleb	29
commemorative coins	30
complete credit card payment schedule	24
complete mortgage payment schedule	23
compounding	3 *
continuous interest	32, 36
corporate bonds	28
coupon bond	28
credit card trap	24

credit cards	2, 24, 25
danger of credit cards	25
Declaration of Independence	36
defacement of currency	38
Denver Mint	37
Department of Treasury	3, 6, 15
Deposit accumulation models	10, 11
Derek, Bo	11
Diner's Club	25, 26
discounting	4 *
Disraeli, Benjamin	41
Dooley, Ken	30
E Pluribus Unum	11
early saving	32, 35
effective rate	18 *
equivalent interest rates	18, 20
extra mortgage payments	32, 37
Federal Reserve	21
Fifty States Quarters Program Act	33
financial tables	6, 14
Fixed Rate Mortgage (FRM)	42
Flynn, Errol	31
Forbes, Malcolm	12
foreclosure	1
fractional exponents	5 *
Franklin, Benjamin	5, 17, 26
Getty, J. Paul	15, 43
government bonds	28
graphics calculators	30
Hamilton, Alexander	17, 20
HIDE ROWS (Excel)	23, 37
high credit card rates	25
Hope, Bob	18, 22
Hubbard, Frank McKinney	2
In God We Trust	16
Independence Hall	2
inflation	7, 32, 40
inflation retirement concerns	32, 40
installment payment formulas	17
installment payments	16 *
interest	3 *
interest portion of credit card payment	24
interest portion of loan payment	16 *
intersect calculator feature	13
introductory credit card rates	25
investment yields	13 *
IRA	27

Jackson, Andrew	18, 19
Jefferson, Thomas	28, 36
Jones, Franklin	13
Kennedy, John	28
Lincoln Memorial Building	10
Lincoln, Abraham	28
linear equation	9
loans	16 *
Locke, John	44
logarithms	4 *
Lorimer, George Horace	8
Lyons, Judson	6
Madoff, Bernard	31
Mansfield, Katherine	1
Master Charge	27
MasterCard	27
maturity value of a bond	28
maxed out (on credit card)	25
meltdown	1
Mencken, H. L.	33
Milligan, Spike	9
monthly payments	21 *
mortgages	1, 22 *
Morton, Azie Taylor	6
multiple deposits	8 *
municipal bonds	28
Murphy, Thomas P.	20
Napier, James	6
negative exponents	4 *
nominal interest rates	18 *
Novus Ordo Seclorum	13
Orben, Robert	17
order of operations (algebra)	1
outstanding principal	16 *
Parker, Dorothy	34
Patriot Bonds	29
payday loans	32, 34
penny doubling	32
pensions	2 *
Philadelphia Mint	37
piggy banks	42
Ponzi scheme	31
present value	12 *
principal portion of credit card payment	24
principal portion of loan payment	16 *
pygg	42
pyramid on $1 bill	13

296

pyramid scheme	31
Raft, George	7
Rand, Ayn	40
Reagan, Ronald	20
refinancing	32, 38
Renard, Jules	35
retirement funds	26, 40
retirement plans	26, 40
Rogers, Will	1
roll-over feature of payday loans	34
Roosevelt, Eleanor	32
Roosevelt, Franklin D.	28
Roth IRA	27
rule of 72	32, 33
San Francisco Mint	37
Sanders, Colonel Harland	19
Sarnoff, Robert W.	38
serial numbers (on U.S. paper money)	1, 24
Series E Bond	29
Series EE Bond	29
simple interest	3
single deposit accumulation	3 *
social security	2
spreadsheets	3 *
symbol $	12
tax deductible item	23, 25
teaser rates	32, 41
torn dollar bill value	40
trapping a solution	5, 9, 11, 30
trial-and-error process	5, 9, 11
true annual rate	18 *
Trump, Donald	39
Turnbull, John	36
Twain, Mark	42
U. S. Treasury	4, 22, 29, 35
U.S. Savings Bonds	29
United States Secret Service	38
Ustinov, Peter	25
Vaughn, Bill	23
Vernon, William	6
Visa	27
Vivian Grey (novel)	41
Washington, George	1, 28
Washington, Martha	5
weight of paper money	14
West Point Mint	37
Whitehorn, Katherine	10

Wilde, Oscar	27
yields	13 *
yields on coupon bonds	30
Youngman, Henny	6, 24